As I read through Brad's bc
quite a few similarities in life. But the similarity that most struck me was the story of the death of Brad's dad. My dad, too, was bigger than life, and the day he died things seemed to stand still. My life and the lives of all our family and friends would never be the same. We had two choices: we could either quit, or we could carry on with even more drive than before.

This book is not so much a book about failing, as we will all fail from time to time, but about getting back up when you have been knocked down. It doesn't matter how many times you have been knocked down as long as you keep getting up. This book shows that God orchestrates every day of your life. I hope you will be encouraged to do more and be more after reading this book.

—Kelley Earnhardt Miller
Vice President, General Manager
JR Motorsports

Brad has the gift of showing how God reaches out to us through both the difficult and the good experiences in life. He shares his amazing life stories (which are so crazy I actually had to verify them) on how we can trust in God as he shapes our lives. I've known Brad for many years, and I can't wait to see the impact his book will have and the lives it will change.

—J. D. Gibbs, President
Joe Gibbs Racing

Brad's book is a reminder to us all of what is truly important in life. The path is indeed narrow, and through Brad's journey, he shares with the reader how easy it is to venture off the path that brings life. He spotlights the everyday struggles that face us and the effort involved in making godly decisions rather than worldly ones. I personally felt that many of Brad's stories mirrored those of my own journey. His book serves as a great reminder to us all that God's timing is perfect.

—Billy Huard
Former NHL Hockey Player

What a book! As I see God's strong hand in Brad's many life challenges, I am inspired to keep going. I've learned from Brad that the lows of life are merely intended for us to build speed in order to reach highs barely imaginable.
Thanks, Brad!

—Robb Williams, Pastor
Verde Baptist Church

Former President Richard M. Nixon once said, "Well, I screwed it up real good, didn't I?" All of us have experienced the sting of failure. Sadly, however, defeat all too often destroys us. Brad didn't let that happen to him. Though his times of failure definitely impacted him, they never defined him.

Brad's book is for anyone who has wrestled with failure, and it provides encouragement for us to never give up. But what I love even more about this book is that it's not a story about Brad's failure as much as it is about God's faithfulness. God is the hero of Brad's story. I'm eager to see how God uses it for the sake of his own name.

—James D. Metsger, Pastor
Renaissance Bible Church

It has been a true blessing in my life to become a brother in Christ with Brad Henry. I have read every book he has written, and I have even lived some of the experiences that he has shared with his readers. This book will prove to be his best work yet, I believe, and I challenge you to read it and pass it on. Brad is truly God's man, and he has answered that call by displaying his talents through writing this book. Awesome job, brother! Keep up the great fight.

—Reed Nettles, CEO
HLG, LLC

Failing Doesn't Mean You're a Failure

How To Turn Failure Into Success

Brad Henry

The Ultimate Decision
P. O. Box 2337
Huntersville, NC 28070
704-907-8396

brad@theultimatedecision.com
www.theultimatedecision.com

Failing Doesn't Mean You're A Failure
The Unexpected Road to Success
by Brad Henry

Printed in the United States of America

ISBN 9781619961531

Some names have been changed by the discretion of Brad Henry.

www.xulonpress.com

Dedication

I dedicate this book to Jesus Christ and Him alone.

I am a sinner, but Jesus, you saw something in me that I didn't see in myself. You saved me when I didn't deserve to be saved. You gave me a gift to write even though I was diagnosed as having dyslexia and ADD. You and you alone deserve all the glory and honor for this book, and I pray that many will come to know you personally through it. I now see that I cannot do anything good in this world unless you are behind it. Thank you, Jesus, for orchestrating every day of my life.

Contents

Foreword

Brad Henry's book *Failing Doesn't Mean You're a Failure* takes its readers through his journey of first living life without God and then living a transformed life with and for God. I think the ultimate goal of this book is for Brad to draw his readers' attention to the fact that God is all-knowing. Romans 3:23 says, "For all have sinned and have fallen short of the glory of God," and God, of course, knows that. But God is also all-loving, and his grace is expressed in 1 John 5:13: "I write these things to you who believe in the name of the Son of God so that you may know that you have eternal life."

As I read Brad's story, I related it easily to my own personal story with its specific circumstances and characters, and I think many readers will do the same. It is only pride that keeps us from realizing God's great sovereignty. Until we realize that we were not created for our own purposes and desires but perfectly created for God's perfect purposes, happiness will be a fleeting and conditional emotion. But when our hearts are truly changed by the presence of the Holy Spirit living inside our hearts, we will no longer feel compelled to stop our immoral ways out of duty and guilt, but we will cease doing them out of love for the permanent joy and peace that Jesus gives us.

If you have never heard the gospel of Jesus Christ, you might want to skip to chapter 32 before you read the rest of the book. Deciding what to do about Jesus truly is the biggest decision of your life. In chapter 31, Brad says, "God knows your heart and your desires. He knows what it takes to force you to slow down. But I have found that when God sees you taking measures to slow down to listen to him, he will move mountains for you." To that I will testify, "Yes, he will!" As Brad so clearly expresses, God can save you and anyone else who wants to be saved, and only Jesus can lead you to finding the purpose you will read about in this book.

Brad and I share a common understanding and love for the sport of triathlon, but thankfully we were brought together by a greater passion: to serve our Lord Jesus in all that we are and all that we do. Thank you, Brad, for asking me to write the foreword of your book. I feel honored to be one of the first to read your God-inspired story of despair, then hope, and ultimately redemption.

I pray that each reader of this book will be inspired to know Jesus in a more personal and intimate way by studying his Word. I also pray that each reader will bless another by choosing to follow in God's ways and then testifying to God's cleansing grace, saving power, and free gift of salvation.

—Sian Welch
Ironman Champion

Acknowledgments

Many people are responsible for the completion of this book. First of all, I want to thank my wife Julie and our two boys, Bryce and Chase, for their love and support. No matter how many times I have seemingly failed you, you have shown me what true love looks like. I love you.

To Mike and Ana: What can I say? Words cannot express my gratitude for your faithfulness.

To Tony and Billy: You are two men who have stayed the course with my family and me in many different ways. Thank you for believing in us.

To Paul and Elaine: Thank you for your wisdom and grace. It will never be forgotten.

To my in-laws, Duke and Carole: Thanks for always coming through when we need you the most.

To my mom: You had every right to be mad at life, but you taught me how to persevere.

To all the prayer warriors and supporters of the Ultimate Decision, my ministry from God: Your prayers and support have lifted my spirits when I was down and have enabled me to keep fighting the good fight. This side of heaven, you will never know how much you mean to me.

Flight 1992

"Worldair 1992, turn left heading 270; descend and maintain 8,000."

"Roger. Out of 9 for 8, Worldair 1992."

I was copilot on this flight with Captain Leo Stutz. He and I had been flying together for the month of December. On this day, we had started in Cincinnati around eleven o'clock that morning, then flown to Detroit, Michigan, and then on to Nashville, Tennessee. It was now nighttime, and we were on the last leg of the journey, heading from Nashville into Indianapolis, Indiana.

"Worldair 1992, turn right heading 300; descend and maintain 6,000 feet and contact Indianapolis Approach on 125.6."

"Roger. Out of 8 for 6, turning right to 300, Worldair 1992."

"Worldair 1992, acknowledge Lima, as there has been moderate to severe icing reported on the approach."

When flying into any airport, pilots use an acronym called ATIS (Automatic Terminal Information System) that gives them the weather, temperature, visibility, wind direction and speed, landing runway information, and any other pertinent facts they need to know for landing. The nonflying pilot will usually work the radio. On this flight into Indianapolis, I was flying and Leo Stutz was operating the radio. Leo was a no-nonsense ex-military fighter pilot, and he had the buzz cut to show it. With him everything was done by the book.

Leo began writing down the information: "Indianapolis weather 22:00 Zulu; temperature 31 degrees, sleet and freezing rain; ceiling 700 feet visibility, ½ mile and fog; winds from 050 at 10 mph; landing runway 5 right; confirm you have Lima."

Leo clicked the mike and said, "Indianapolis Approach, Worldair 1992 has Lima."

"Roger, Worldair 1992. Turn right, heading 360; descend and maintain 3,000 feet."

"Roger. Worldair 1992 descending to 3,000 feet, turning right to 360."

As we continued our descent, I spoke to my partner. "What a day, Leo! I can't wait to get to the hotel tonight and throw back a few beers."

"I know, Brad," Leo laughed. "Maybe we can get those same ladies to buy drinks for us like they did the last time we were there."

"Okay, that sounds great," I agreed, "but let's get this thing on the ground and call it a day."

"Worldair 1992," said Indianapolis Approach, "maintain 3,000 and turn right heading 020, and intercept the localizer for runway 5 right."

"Roger. Worldair 1992 maintaining 3,000 feet, turning right 020 to intercept localizer for 5 right,"

"Leo, looks like we are picking up quite a bit of ice," I commented. "I can feel it in the flight controls. Make sure we get the landing checklist right."

"Ready, Brad."

"Yeah, go ahead, Leo."

"Circuit breakers in, check. Fuel pumps on, check. Landing lights on, check. Anti-ice on, check."

"Okay, that's good for now, till we get over the outer marker. Localizer alive; roger, Leo, turning to intercept." said Brad.

"Brad, look at that windshield wiper," Leo pointed out to me. Since it is almost impossible to see the wing of an airplane, a pilot has to use other clues to determine how much ice the outer surfaces are picking up. The mechanism that provides the hinge for the wiper blade always accumulates ice, so it is a good clue to outside conditions. Also, a decrease in the aircraft's speed is a good indicator of ice on the wings.

Within a matter of minutes, the ice on the hinge of the wiper looked to be the size of a baseball. "Brad, I am cycling the boots and the props," Leo said.

Pow—pow—pow!

"Man, what was that Leo?" I asked, startled and concerned. We had accumulated a great deal of ice on the props, and as the props started to heat, they hurtled the ice against the side of the aircraft, thus producing the jarring sound.

"Keep up the heading and altitude, Brad. I am going to talk to the passengers and tell them what they just heard."

"Roger, Leo."

"Ladies and gentleman, this is your captain speaking," soothed Leo. "Sorry for the loud noise you just heard. That was our props heating up and throwing off the excess ice. We should be on the ground in about seven minutes. Flight attendants, prepare for arrival."

"Worldair 1992, descend and maintain 2,500 feet; slow to 150 knots. You are four miles from the outer marker. Contact Indianapolis Tower 119.9."

"Roger, Approach. Worldair descending to 2,500; contacting Tower 119.9. Indianapolis Tower, Worldair 1992 down to 2,500 feet, slowing to 150 knots."

"Roger. Worldair 1992 cleared to land runway 5 right."

"Roger. Indianapolis Tower cleared to land runway 5 right."

"Okay, Brad," spoke up Leo, "landing checklist: flaps 15 degrees; reduce to 130 knots."

"Full power, Leo! I need full power," I yelled frantically.

"Brad, what the hell are you doing?"

"More power—more power, Leo! I can't control this damn thing . . ."

....

"Ether, it looks like we have Brad Henry in our grasp tonight. In just a few moments, he will be with us to spend eternity in hell."

"Now Satan, I know you have power, but don't you have to check with Jesus first?" the evil minion answered.

"Don't ever say that name in front of me again, Ether!" demanded his master. "Do you understand me?"

"Yes, yes, Master, and please let go of my throat," pleaded the demon.

"This time I am going to kill Brad and bring some passengers along for the ride. How does that sound, Ether?"

"Sounds good to me, Prince. Brad has always been a womanizer, with his mind fixed on drinking and women. The ice I am putting on this plane will bring them down in just a few seconds. Now tell me again why all the other people on the plane are going to hell tonight too?" questioned Ether.

"Well, there are a few on board tonight who have professed their faith for J, and I can't touch them," Satan admitted, decidedly disappointed. "They will experience an earthly death tonight, but sorry to say, they will go to heaven. We have lost those, but we haven't lost Brad. I have kept him busy with his lustful heart, idolatry of money and excessive drinking. He is still mad at J for taking his father at a young age," he gloated.

"Yeah, that was a good one, Prince. You made him think that J took his dad when it was really you who killed his father," Ether admiringly stated.

....

It was that time of the day when the angels come to get their orders from Jesus. Addressing them, he said, "As you know, Satan and his demons are trying to destroy those who have not accepted my free gift of salvation. They have no power, but we still have to fight them for the souls hanging in the balance."

The mighty Gabriel arose and said, "Tonight Satan is trying to destroy one you have destined to inherit salvation on November 29, 1992."

"Yes, I know Gabriel," Jesus answered. "I created Brad Henry and have been with him every day of his life. He thinks I am the one who killed his father, and he doesn't want anything to do with me. But I know his heart, and I will not let Satan have him tonight or any other night for that matter. Brad is mine, but he must suffer many trials because he is a prideful person."

Jesus continued, "Over the years, I have been very amused at times with Brad's life. He has failed in so many things so many times you would think he would have asked me for help by now."

"So why hasn't he?" Gabriel asked.

Not immediately answering the question, Jesus replied, "I want people to come to me because they love me, not because I force myself upon them. For right now, Gabriel, send angels to hold up that aircraft and bring it back to the ground safely. Then get me Satan."

"Jesus, the angels have been dispatched," said Gabriel, acting immediately upon the Savior's orders. "I, too, am interested in the life of Brad Henry. If I may ask again, why is it taking him so long to make up his mind about you?"

"You know, Gabriel, that a day is like a thousand years to me, and a thousand years is like a day. From the beginning of time, man has been trying to figure me out. He spends countless hours trying to understand what will never be revealed on this side of heaven. I have left a place in each man's heart that is reserved just for me. They try to put boats, cars, money, lust, power, and other things in that hole, but it is only I, Gabriel, who can fill that empty void."

"So why don't you stop Brad's pain and make him see you for who you are?" the puzzled angel asked.

"All in good time, Gabriel, all in good time."

Good Times

We were a close-knit family—my mom, my dad, my brother Jeff, and I. My father was a doctor in Sarver, Pennsylvania, a small town in the western part of the state, and served many in the community. We took vacations together as a family, and my brother and I had a great life growing up, never wanting for anything.

My dad was a very religious man and tried to get my brother Jeff and me to church whenever the doors were open. He also encouraged my talent in athletics and was my Little League baseball coach. The two of us were inseparable. With Dad as my coach, I hit more home runs than anyone else, pitched a no-hitter, and even hit three home runs in one game. Dad was proud of me, and I tried to do whatever I could to impress him.

My dad was also an avid and meticulous hunter. He wanted me to experience the same passion he felt for the sport, but I was more into comfort than roughing it. On one trip we took together, it was the opening day of buck season and my first time to go deer hunting. The temperature was about five degrees above zero, and ice mixed with snow covered the ground. Dad and I were sitting on a log, trying to be as quiet as possible. Dad had some chew in his mouth and occasionally needed to spit. He would look first with his eyes to the left and then to the right to make sure there were no deer around. Then he would turn his head to the right and then to the left, slowly drop his head, and spit as quietly as he could.

Now that was what was happening on his side of the log. On my side of the log, I was so cold I could think of nothing else, even though I had a hand warmer in each pocket. I also had a stash of about five candy bars in my pockets, and not really caring about the deer, I just dug into my sweet treats. Had there been any deer around for five miles, they would have certainly heard my commotion. Despite all the work my dad was doing to keep quiet, I gave our position away in just a few seconds. Dad just rolled his eyes at me and threw me a warning look to be quiet.

I have always been a shoot-then-aim type of person, and that was never more apparent than that day when I was hunting with my dad. All of a sudden, a big buck appeared about seventy-five yards in front us. My dad motioned for me to take the shot. I put the crosshairs of the scope on the buck's shoulder where his heart would be. Then I tried to shoot, but my gun was jammed. Well to be honest with you it wasn't jammed. To put a shell in the chamber of the gun to shoot you have to pull back on the bolt and then inject the shell into the barrel. Without knowing it, I just started pushing and pulling the bolt and ended up ejecting all my shells. I only pulled the trigger when I was out of shells, Oh the horror!! Again my dad just rolled his eyes.

Later that day, we were riding in our jeep on a logging road to another part of the forest. All of a sudden, about eight deer sprinted across the road. I jumped out of the jeep and started firing away. After I got back into the jeep, my dad said, "Brad, do you know where those shots were hitting? About twenty feet up in the trees." Buck fever is bad, and I had it bad. Needless to say, my first exposure to deer hunting was not the experience either my dad or I was looking for. Dad was not too pleased with my hunting ability, and I was soon ready to go home to a warm house.

My dad was a strict disciplinarian, but you could tell he really cared about his children. He wouldn't say it outright, but you could tell he was proud of his boys. My brother Jeff got straight As, and I was lucky to get Cs. My mom and dad didn't know it at the time, but I was dyslexic and also had ADD. It was not a common diagnosis in 1966. No matter how hard I concentrated or tried to read, I struggled to comprehend. This gave my dad fits. How could a doctor's son be so dumb?

Each night Jeff and I had to study for an hour. After the hour was over, Dad would ask us questions about our lessons. Jeff always had the right answers. I, on the other hand, could never get anything right. I couldn't concentrate; it was horrible. I would read two sentences and then start daydreaming and have to read the same sentences all over again. I would soon give up, resigned to suffering the consequences. There was no use trying, I eventually concluded.

My dad sent me for an IQ test to see what was wrong with my learning ability. However, the day of the test was also the day of the Halloween parade at school. Well, you can imagine where my mind was. The report that came back a few weeks later said I was not dumb but had been more interested in getting back to the Halloween parade at school than in taking the test. At least I excelled in baseball and football, which drew praise from my dad and also gave me the self-confidence I would need to get through some rough patches in my life.

In my seventh-grade English class, we had to write an essay about our best friend. Everyone wrote about a classmate—that is, everyone except me. I wrote about my dad. Without my knowing it, my teacher mailed the letter to him. It would be years later before I discovered what my teacher had done. My dad never let on that he got that letter, but from what my mother later said, it meant the world to him.

………………

"Ether, gather some of the other demons; we have a problem."

"What is it, Master?"

"I don't want this Brad Henry to accept J's nonsense of salvation. We need to get him mad at J. The next time we are called to a meeting with J, I will bring this little boy up to him."

....

“Satan, what is it you want?” demanded Jesus.

“Brad Henry has had a privileged life, and he goes to church and praises your name because everything has been great in his life. But take his entire family from him and he will run to me—not you, J.”

“Satan, I have plans that you are not privy to. I will show you what a man’s heart is made of. You may take his father, but you cannot touch Brad, his brother, or his mother.”

Gates of Hell Opened

My mom and dad had just returned from the Cleveland Clinic, where my dad was tested for heart problems. The doctors concluded that the problem was his gall bladder. My mom later said that he slammed his fist onto the table and insisted to the doctors that it was his heart. Even if he had been diagnosed with heart problems, the heart bypass had not yet been perfected in 1969. There was not much they could do, except what they did: they told him to go home and take it easy for a couple of months.

"Hi, boys. How did school go today?" asked Dad.

Well, as usual, Jeff had done great in his classes, but all I could say was, "I did really good at football practice, Dad. Do you want to go out and throw the football tonight?"

"Brad, I'm a little tired. Why don't you and Jeff go out and play for a while? Then come back and we'll watch some TV together."

"Okay, that sounds good, Dad."

After throwing the football with our friends in the neighborhood, Jeff and I went back inside our house. We ended up in the spare bedroom, watching a TV show with Dad and our dog Sam. Jeff and I were lying on one bed, and Dad occupied the other. It was going to be a fun night of TV and popcorn, with Sam looking for a few bites to be thrown his way.

All of a sudden, Dad started breathing quite hard and slumped over in the bed. Jeff and I ran out of the room, yelling for Mom.

Panic-stricken, Mom quickly responded as she came running, "What's the matter?"

Running back into the spare room with her, Jeff and I kept yelling, "Dad is dying! Dad is dying! Mom, what's going on? Mom, what is going on? Tell us, tell us, is Dad going to be all right? What is going on, Mom?" Our screams that night were horrific.

My dad's partner in his medical practice rushed over to the house, but it was no use. At the age of thirty-four, my dad died of a massive heart attack. Our household would never be the same. Our rock and our future security were now gone forever.

That night Jeff and I slept together in our parents' room. The doctor gave us a sedative, but we still cried ourselves to sleep, hoping this was just a bad dream. But when we woke up in the morning, the horrible nightmare was really no nightmare at all. It was the reality we would now have to face for the rest of our lives. The house was chaotic with everyone coming and going, giving food and offering condolences. This did little to ease our pain, though.

....

Gabriel asked Jesus, "Why did you take Boyd Henry when he could have done so much good as a doctor and been there to raise his boys?"

"As you know, Gabriel, man has always put too much hope in this life on earth. It is through trials that man comes to me, not through a lifestyle of luxury. When man has too many material things, he doesn't think he needs me. I want people to hunger after me. I also want people to understand that earth is not man's real home; it is either heaven or hell. Life on earth is just a brief blip on the screen of eternity."

"So is that why you allowed Dr. Henry to be taken tonight?"

"Gabriel, Dr. Henry was a believer. He put his faith and trust in me at a young age. He is in heaven. But what Satan does not know is that through the death of Dr. Henry, a call has gone forth for people to wake up from their slumber. Many will come to know me through the death of this man and the story I am going to build through the life of his son Brad."

Gabriel replied, "I know you can do all things, but Brad is very far from you now, Jesus."

"All in good time, Gabriel, all in good time."

....

"Ahhhhhh! Now I see everything and why it has happened, Jesus!" exclaimed Boyd Henry. "All my questions have been answered in one second. Words cannot describe what I feel and see. The music is awesome, and it sounds like the roar of rushing water. Mom, Dad, Beebe, David, Tom, Bill—it seems as if it has been only a day since I last saw you."

"It seems like only a day since we have seen you too, Boyd," his mom sweetly replied. "We have so much to talk about and to show you. The glory of the Lord gives heaven light, so there is no night. We can talk, sing praises, and have wonderful conversation for eternity. Let me take you to your mansion."

"That's right, Mom—a mansion. Wow!"

....

"Jesus, does Boyd Henry remember his death or life on earth?"

"No one will ever remember their earthly life or death, Gabriel. Since no pain or suffering exists in heaven, Boyd will not remember his family left on earth or his life there. He will, however, remember in an instance all the good of everyone he sees in heaven from his past life on earth."

"Won't Boyd miss his wife and children?" Gabriel asked.

"No, Gabriel, as that would be pain. But when he finally sees his family again, he will recognize them, and it will seem like only a day that they have been apart. The pain is for those left behind, Gabriel, and I allow that pain for two reasons. One is so I can comfort the brokenhearted, and the other is that I want people to think about their eternal destination.

"I want Boyd's death to shake people out of their slumber. If people on earth would only realize how awesome their real home is, they would rejoice, not cry, when someone dies and goes to heaven. But most people would rather live in their sin on earth because that is all they know. Most would rather live by sight than by faith.

"Boyd never has to worry again about pain of any kind. Now we have to help the rest of the Henrys move on, as they are not sealed for eternity—at least not yet."

………………

"Do I have to go to the funeral home, Mom?" I asked apprehensively.

"Yes, Brad. Many people are coming, and you would regret it if you did not go," explained Mom.

As we drove to the funeral home, neither Mom nor Jeff nor I said a word. Nothing could be said that would ease the horrible pain.

"Mrs. Henry," said the funeral director kindly, "I am so sorry for your loss. Are you ready?" I remember thinking, *What does he mean, "Are you ready?"*

"Come this way," he directed as he gently guided Mom, Jeff, and me to a room with a drawn curtain. Mom grabbed Jeff's and my hands, and the curtain was drawn back. There was my daddy. I wanted to tell him how much we were hurting and to ask him to please hold us. Why couldn't he make it better just one more time? Why did he have to go? How would we go on? How . . . how . . . how?

My dad had been the physician for our town's football team, so all the players came to pay their respects. I was on the freshman team, and my coaches came over to talk to me. It was all I could do to stand up. I couldn't hold onto them like I could hold onto my daddy.

Dad, this is so tough. I don't know how I can do this. I don't know how I can go on. I miss you so much. Please wake up and help me! Desperate thoughts swirled through my mind.

The day of the funeral service and burial arrived, a freezing February day in western Pennsylvania. I don't remember much of the funeral, since I tried to think about other things. But what I will always remember is what happened when the service was over. We all walked out of the church, but as I exited through the door, it dawned on me that I would never see my dad's face again. I ran back inside before the casket was shut and took a long, hard look at my dad; I couldn't stop crying. This moment is burned into my memory some forty years later as I am writing this. After a few minutes of gazing upon my dad, I knew that nothing would change and that I was now on my own. But those last few minutes with my dad were precious, and I am glad I went back.

It was a few months before I could even go back into the bedroom where Dad died. The worst part of each day was at dinner. As a family, we had always eaten dinner together, but Dad's empty chair was like the proverbial elephant in the room. We all knew it was there, but no one wanted to say anything. So we each buried our pain, even buried our feelings for one another. My mother, brother, and I loved each other deeply, but we certainly didn't want to feel this kind of pain ever again. In a profound way, a big part of each one of us died with Dad. We would never be the same again—nothing would ever be the same.

A Gift of God Revealed

"Gabriel, I am going to give Brad the ability to be a strong long-distance runner. This will help ease the pain from the death of his dad and give him something positive to focus on. I am also going to send him to new surroundings. I am going to send him to the prep school his father attended. It will provide some quiet time for him to build up his confidence."

………………..

In the fall of 1974, I was fourteen years old and enrolled as a freshman at the Kiski School in Saltsburg, Pennsylvania. Kiski was a boys' boarding school located in the country far from anything else. We only got to go home once every six to eight weeks. Enrolling at Kiski gave me the opportunity to get away from my normal surroundings that reminded me of my dad and provided a chance for me to find my own way.

Each dorm on campus had a teacher who lived in the dorm and a prefect, an upperclassman who helped maintain discipline. School started each morning at eight o'clock. Each student had to wear a sports coat and tie to classes, to breakfast, and to lunch; then a white shirt, coat, and tie to dinner. From seven thirty to nine thirty each evening, we were supposed to study in our rooms, with no talking allowed. At a quarter past ten, it was lights out, and again no talking.

My first evening at Kiski was quite memorable. My roommate, who had taken the upper bunk, leaned over as I lay on the bottom bunk and said, "Hey, Brad, listen to this." Remember how I said no talking? Well, Chuck didn't care about that and proceeded to lean over his bed and pass gas for at least ten seconds. He kept demonstrating his unique talent until I was screaming with laughter.

All of a sudden, the light in our room flipped on. The prefect sternly reprimanded us, saying, "You think it's funny being here? Well, come on out into the hall." We were dressed only in our underwear, but that didn't matter, since Kiski's was an all-boys school. "Okay, guys," the prefect continued, "since you have so much energy to spare, you are going to push a penny down the hall with your nose."

For the next thirty minutes, Chuck and I each pushed a penny down the hall while keeping our hands behind our backs. After that the prefect said, "When we say no talking we mean it, understand?"

"Yes, sir!" we both practically shouted. We went to bed dirty, but tired. We never talked again after lights out.

At the school, I started as a quarterback on the JV squad and lettered on the varsity football team as a receiver. But at a game in Buffalo, New York, I suffered a severe concussion, and my football career was over. Fortunately, I attracted the headmaster's attention, though, during preseason football. During preseason camp we had to run a mile to see what shape we were in. That is when the headmaster noticed my running ability. My time was not far off the school record. That training run would set me on a whole new course in my life.

"Brad, I didn't know you could run like that!" exclaimed Mr. Pidgeon, the headmaster at Kiski School and also the track and cross-country coach. Through the preseason in track the following year, it became apparent that I had indeed been given a talent for running. One night in the library, Mr. Pidgeon pulled me aside and said something that would change my life. "Brad, you are going to be the best runner this school has ever seen," he declared.

"Okay, guys, today we are going to do quarters on the track," said Mr. Pidgeon one day at practice. We all groaned whenever Mr. Pidgeon announced our workout plan. It always seemed insurmountable because we knew just how painful it was going to be. The workout for this day would be a mile jog, followed by ten to fifteen quarters. A quarter was one lap around the track, and the workout today would consist of running each lap at just under 5 minute per mile pcae. Each lap would be run in approximately sixty-five-to-seventy-seconds per lap followed by a jog without stopping. To run five minute mile pace would be 75 second quarters so this pace was pretty fast. When we reached the starting line after the jog, Mr. Pidgeon would yell, "Yoooooo!" and we would run a fast lap again. Then the workout would end with another easy mile jog to loosen up.

Just as we finished the workout and were about to head for the locker room, Mr. Pidgeon yelled in a booming voice, "We aren't done yet, boys." Okay, what now? Mr. Pidgeon was a great coach and even better motivator. He had to be to get us to run like this.

"I know you are tired, men, but that is what separates the men from the boys. Let's go to the ski slope, and we'll run it three times!" he encouraged. I was already tired; I had run five miles in the morning, and now in the afternoon, we had just run thirty-eight laps around the track at a good pace. Now our coach wanted us to tackle the ski slope. At least, we were going to be in better shape than most other teams.

I did not know it at the time, but I was starting to develop an unhealthy attitude toward running, an attitude that would eventually bleed into every other part of my life. If I ran a quarter in seventy seconds, then I felt I had to run the next in sixty-nine seconds. It always had to be better. I would train so hard that I would eventually throw up. Then if I didn't throw up, I thought I wasn't training hard enough. So for the next three years, I threw up at least once almost every day. I pushed and pushed myself so hard that some forty years later, I still have problems with my stomach.

I eventually broke the record for the cross-country course, but even though I won the race by a large margin, I needed to break the course record again to feel any satisfaction. I ended up breaking the course record over six times, but the whole thing eventually became a huge burden. The fun of winning was wearing off unless I broke a record in the process.

Our first track meet of the year was held at Indiana, Pennsylvania. All the schools in the county were at the meet. Our school record for the two-mile was nine minutes and forty-nine seconds. Mr. Pidgeon walked up to me and said, "Okay, son, let's start out easy at seventy-five-second quarters for the first mile, and then we we'll see how you feel." Seventy-five-second quarters is a five-minute- mile pace.

The gun went off, and many of the other contenders surged past me, but I knew from experience to let them go. Most would fall back after a few laps. "Remember, just run your pace," Mr. Pidgeon would always say. At the end of the first mile, Mr. Pidgeon looked at his watch; my mile time was four minutes and fifty-eight seconds. What pleased Mr. Pidgeon even more, though, was that I was smiling. Actually, I felt great.

"Okay, Brad, do what you can. You have four laps left," he called. The next two laps I ran in seventy-three seconds each, and then I ran the seventh lap in seventy-two seconds. The school record was now in reach. I needed to run another seventy-two-second lap in order to break the record in the two-mile. With 220 yards to go, my adrenaline kicked in, and I found another gear. My final lap was run in only sixty-three seconds, shattering the school record in a time of nine minutes and thirty-eight seconds. Mr. Pidgeon was correct: I was now considered the best runner in the school's hundred-year history.

I eventually broke this record numerous times and in the fall received an offer to run for the University of Maryland. When I got to Maryland, however, things were different. At Kiski I had won races by minutes over my closest competitors. Now I could hear everyone breathing down my neck, and I certainly wasn't first. But God had given me a gift to help with the pain of losing my dad, even though at the time I refused to give God credit for anything.

In one particular workout in Maryland, we were to warm up with a mile jog then do five one-mile repeats on the track at a pace of four minutes and forty five seconds or better per mile. The large stadium at the University of Maryland held over thirty thousand people for football games and was a cool place to train. My last mile of the five-mile workout was run in four minutes and seventeen seconds. This was faster than my record-breaking performance a year earlier at Kiski.

My best performance came in a meet against North Carolina at Chapel Hill when I ran 5.8 miles in just a little more than twenty-eight minutes, which was close to a sub-thirty-minute six-mile. Yes, the gift was larger than I had imagined.

That year in cross-country, we won the Atlantic Coast Conference (ACC) championship, and I was one of two freshmen to letter on the varsity team. Things were good, and I was in control of my destiny. God could take my dad, but running was mine.

....

"Yes, Gabriel?" Jesus quietly asked, aware that the angel was waiting to ask a question.

"If I may ask, Lord, what is troubling you?"

"I knew this day would happen, Gabriel. Brad Henry has made an idol out of running. He is consumed by this sport. His heart was already far from me when his dad died, but now it is even farther. I love Brad, but in order for him to come to repentance, he needs my discipline. Running will only take him deeper into himself and cause him to become even more self-absorbed. Brad now thinks he is the one who is great at running. He has no clue that it is I, Gabriel, who gave him that gift. It is now time to end Brad's running career. I have great plans for Brad, but they will take time."

....

"Satan, why have you not been doing anything with Brad Henry lately? I thought you were out to destroy him?"

"I am, but he is doing a great job himself of self-destruction. His success at running has made him prideful and arrogant. He thinks he is the best and is invincible. I am putting in his path many women to disrupt his life."

"How will you do that, Prince?"

"Brad needs love, and he is going to find it one way or another. I have a sense that his running career is coming to an end. I need to keep him occupied with a lustful heart. Once that is in place, destruction is not far away."

………………

I spent four years at Kiski, an all-boys school. Since we did not have any girls at our school, that meant we had no dating. We did have a dance with an all-girls school once every four months, but that was awkward, to say the least. When I arrived at the University of Maryland, I went from not having any girls close by to being surrounded by ten thousand girls on campus.

In this new environment, I started to wonder why I was devoting all this hard training to running when I could be dating and partying. I talked to my friend Jamie, who was on the cross-country team with me, and told him, "I don't know what has come over me, Jamie, but I don't have the desire to run anymore. I just want to party and find some women to go out with. I don't even have a desire to study." All the discipline that had served me so well as an anchor in my life seemed to be vanishing. In a scary way, I started to feel like my life was unraveling at the seams.

Where Is My Security?

I finally got a date one night. A friend of mine said that the girl I was going out with loved to smoke grass. I had never smoked it, but this was the 1970s and it was easy to find marijuana. My friend gave me a joint for that night, and I tucked it into my pocket and went off to a party with this girl.

After the party, I casually offered the joint to my date. She said she didn't feel like smoking and that I should smoke it. So I lit up and had smoked half the joint when something very odd started to happen. From my running, I knew how to check my pulse quickly. Now I checked my pulse and realized it was over 250 beats per minute. I could hardly see the road to drive. I would later find out that I was allergic to the THC in marijuana.

At the age of nineteen, I was rushed to the emergency room of the hospital, where I was admitted to spend the next few days in coronary care. Going from being one of the top runners in the country to being bedridden was a blow to my ego. I started to wonder if my life and health were taking me down the same road as my dad's. Was I going to have a heart attack and die like he did? I couldn't get the tormenting thought out of my mind. The memories were haunting me each waking hour.

....

"Prince, why didn't you take Brad to hell tonight? It was the perfect opportunity."

"I tried, Ether, I tried. But as you know, God puts limits on what I can and can't do. I got Brad's heart rate up to where any normal person's heart would have exploded, but God had him in good shape and protected him. Brad will suffer for the rest of his life for this, though. But at least he is mad at God again, Ether. We will still have our chance."

....

My entire endocrine system was now out of balance. My heart raced for over two hours at this furious speed. It damaged my adrenal glands, and I would suffer panic attacks for the next ten years and be required to take medication for the rest of my earthly life.

When I spoke to my physician, Dr. Willis, he said, "You are one lucky man to still be alive after that episode. I am afraid your competitive running days have come to an end. Your body does not have the stamina and your heart will never give you the output needed to be a great runner. I am afraid you will just have to be like the rest of us, content to jog for the rest of your life."

Okay, God, what are you trying to do to me? You have my attention, but what do you want me to do? Do you want me to work for a living, get married, become rich. Will I be blessed with riches? Are you going to take pity on me for killing my dad? Confusion and anxiety rolled relentlessly through my mind.

....

"Why doesn't Brad get what you are trying to do?" asked Gabriel.

"Brad's heart is hardened, Gabriel, and it will take many years of trials for his heart to be softened. If I softened his heart now, he would not learn all the lessons that I need to teach him through his trials.

"As you know, Gabriel, man doesn't learn to love me through a prosperous lifestyle with no trials. Man learns to love me when he has nowhere else to turn. He learns to love me when he discovers no one else will love him. Brad will love me when he looks back over his life and realizes that I never left his side though he left mine for twenty years.

"I have allowed Satan to tempt Brad in many areas of his life that will take him into many dark holes. But the dark holes that Satan means for evil and destruction will be the very darkness that will turn Brad to the light."

"When will this happen?" asked Gabriel.

"A short time in heaven, but a long time on earth," replied Jesus.

"How are the two time frames different, Lord?"

"In heaven," Jesus explained, "a day is like a thousand years, and a thousand years is like a day. When Brad sees his father again in heaven, it will seem like only seconds since they last saw each other. They will not remember the pain of death. I, along with everyone else in heaven, Gabriel, am not bound by time, but people on earth are. That is why there is always groaning in the midst of a trial on earth."

I'm Shot

I felt like I needed to get home to Pennsylvania and regroup. No matter where you are in the world, there is always a certain place that comes to mind when you hear the word *home*. Even though I didn't like to hunt that much, it was a familiar hobby that evoked thoughts of my dad. I felt closer to him when I was hunting and imagined him smiling upon me as I carried on his tradition.

My best friend Bill and I decided to go rabbit hunting over Thanksgiving weekend. Bill had a beagle that he used for rabbit hunting. A beagle can easily detect the scent of a rabbit and then slowly chases the animal while continuing to bark. When chased, rabbits always come back to the spot where they started.

On that particular outing, Bill and I were separated in thick brush and didn't realize that we were getting closer and closer to each other. Bill's beagle suddenly started barking, and we knew he had found a rabbit. We stopped in our tracks, waiting for the rabbit's sure return. We heard the dog's barking grow faint and then grow louder. This meant that the rabbit was getting closer to us.

All of a sudden—kaboom! Bill was just about twenty feet behind me when he shot. All of a sudden, a severe burning coursed through my legs. Then it hit me—Bill had shot me! I sat down on a stump and pulled up my pants legs, observing where the buckshot had entered my calves. Bill sauntered over to me, holding what was left of a rabbit, and said, "I got it."

I calmly answered, "Yeah, Bill, you got me too." Bill turned white. I had always heard that the guy who shoots someone is usually the one who goes berserk, not the guy that gets shot. I hoped Bill would maintain his composure, because I knew I needed him to help me get out of the woods and to the doctor.

Well, after an hour of the doctor's digging out as much buckshot as he could, he decided to leave the rest in. The doctor joked that I would just be a couple of ounces heavier the rest of my life. Why were weird things always happening to me? I concluded it must be an odd coincidence. As a matter of fact, I have had a lot of odd coincidences in life, some that I would come to remember years later.

…………………

"What happened, Ether?" bellowed Satan.

"Prince, I had Bill's gun directed at Brad's back. He was close enough that the gun blast should have hit Brad's heart from behind and killed him. All I know is that I saw a white flash and then the gun tipped down."

"Well, Brad isn't the one praying for protection," mused Satan, "so find out why an angel showed up. We can't miss chances like this, Ether."

…………………

"I feel that Satan wants to find out who is praying for Brad," Gabriel remarked to Jesus.

"No one is at this point, Gabriel," Jesus answered, "but I have control over Satan and his demons. I have told Satan that he can put Brad through trials and pain, but he is not to kill him."

"So why does he keep trying?" asked Gabriel.

"Satan still thinks he has dominion over me," explained Jesus, "but that is the reason he was banished from heaven.

He wants power and control, and he will do whatever it takes to try to destroy the saints. We are at battle every day, Gabriel. It was good that you brought Michael in to fight on this one. Satan wanted Brad dead."

Jesus continued, "When I rose from the dead, I defeated Satan completely. He has no power over me, but he tries to get to me by harming the ones I love."

"If Brad only knew how much you protected and loved him, he would cry out to you. Why don't you tell him?"

"Gabriel, it is not Brad's time yet. There are many things in his life that need refining. Satan has to have his way with Brad for a while longer before he finally sees the truth. I hope for Brad's sake that it is sooner rather than later."

As Jesus spoke, a look of sorrow crossed his face. He knew what Brad was going to have to go through before coming to repentance, and it was not going to be easy.

Watching Life and Death

I left the University of Maryland and enrolled in a college back home in Pennsylvania. I decided to become a doctor like my dad. I took premed classes during the day and worked for the ambulance service in town when I had time.

"Brad, we have a call about an accident involving two cars. You and John will go." Those were our orders that day.

"I'll drive," said John in response to the order. John had been with the ambulance service for many years, but I had been there only a couple of months. John turned on the siren, and we were off.

"It's amazing, Brad, how sometimes car accidents that look so deadly aren't, and others that look like no one would be hurt are disasters. Let's hope on this one that everyone is okay."

John continued speaking as we approached the scene of the accident. "Should be right around the next bend," he remarked. "Okay, looks like the state police are already here," he observed. "Brad, it looks like someone is pinned under the car; you take her and I'll check the people inside the car."

I followed my partner's direction but was soon at a loss concerning how to proceed. "John, come over here. What should I do?" I called.

"Nothing, Brad, just stay with her," said John.

The woman had been thrown from the vehicle, and the car had run over her head. It was as though her face had been sawed in two. The most horrible part was the bubbles coming from where her mouth should have been. This meant she was still breathing, but it was obvious she wouldn't last long. I stayed with her as the bubbles became less frequent and then stopped. I pulled a sheet over her head and moved on to the next person.

This next person had also been thrown from the car. She was at least seventy years old and was pinned under the vehicle. Her knee was fractured and bent so badly that the top of her toes was touching her hip. The woman was actually speaking with us and remained very coherent. We got her into the ambulance and sped off toward the hospital. She must have had massive internal damage, though, as she died on the way to the hospital.

"John, this is all so weird," I commented afterwards.

"What do you mean, Brad?" he asked, puzzled by my statement.

"Where was God in all this? What did this family do wrong to deserve this kind of punishment?" I asked somewhat bitterly.

"I know, Brad, but try not to think about it. If you let death get to you, then you will be a mess," John advised.

"I know you're right, John, but it still doesn't make sense."

"Maybe we're just not meant to figure out some things," said John.

"Okay, John, don't get religious on me," I protested. Changing the subject, I suggested, "Let's get a beer and forget about this day."

"Okay, we can agree on that," John laughingly replied.

There wasn't much training back in 1976 for being an ambulance driver. We did take CPR and learn some basic first aid techniques. Though I tried administering CPR to a couple of people, I was not successful in my attempts.

Since I was taking premed classes, I spent a great deal of time in the laboratory. On one occasion, we had to put a rat to sleep with ether and then operate on the creature. Hopefully, the rat would survive. To get the rat to sleep, we had to first apply ether to a piece of cotton attached to the top lid of a big glass jar and then place the rat into the jar. Then we had to screw the lid on and wait for the ether to take effect.

Well, I did as I was supposed to do, but then my ADD kicked in and I noticed a pretty girl down at the end of the table. I walked over to talk to her, forgetting all about my rat. After a few minutes, it dawned on me that my rat was still in the jar. I ran over to the poor little guy, but he was so dead that ether was spilling out of his mouth and nostrils.

Then I thought of something to try to revive him. I took the rat out of the jar and laid him on his back. I removed the needle to the syringe and injected a couple of bursts of air into the rat's mouth. Then I gently pushed on his chest with my index finger and followed this with a couple of more shots of air. All of a sudden, the creature's little chest bounced and his heart starting beating again. I couldn't believe it. It wasn't much consolation that I could save a rat but not a human being.

Over the next two years as I worked with the ambulance service, I witnessed everything from suicide to murder. One day an elderly woman who had been lonely too long poured gasoline on herself and then lit the match to end her life. After that terrible incident, I couldn't take any more and went back to school full-time.

While I was in college, I worked full-time in the summer at a reform school. Each home at the facility housed approximately ten guys ranging in age from thirteen to seventeen. As counselors, we had night duty in these homes. On one occasion, the head of the camp asked if any of us counselors knew how to ride a horse. I volunteered that I had once been a guest at a dude ranch. Before I could say another word, the job was mine. Little did they know that my experience on a horse was restricted to a week's vacation at a dude ranch when I was but eleven years old.

Well, I guess I'd better learn how to ride, I thought wryly. I asked another counselor to go out riding with me. He agreed to accompany me, so we saddled a couple of horses and took them out on the property. As we approached the entrance to a large wheat field and started the ascent to the top of the hill, my horse turned around and starting running downhill towards a barbed wire fence. This horse—this messenger of Satan—knew what he was doing. I managed, however, to get my left foot out of the stirrup and then backed my right foot out as well. I quickly jumped off the horse and slid down the wheat field backwards.

At twenty years of age, I was still quite limber, so I got up and walked back to my horse, which I am sure was laughing to himself. The guy I was with offered some very important information about horses, since I obviously knew nothing. "You have to show him who's boss," he said. "Smack him on the nose and he'll get the message."

Well, I pulled back and hit that horse's nose so hard that snot flew in all directions. Now I knew I was in control. Quite cocky, I mounted the beast again. We approached the same spot on the top of the hill, and—oh no!—he turned around and *really* took off downhill towards the barbed wire fence. I had to eject the same way as I had done before.

I admitted to the horse that he had won and walked him a mile back to the barn. I am sure he had a great laugh with all his buddies back at the barn. Oh, I told the head counselor to find someone else to teach riding.

On my twenty-first birthday, a few friends surprised me by taking me to a bar for free drinks. One friend told me that I could drink all I wanted and he would help me drive home. Our return trip involved driving on the interstate for about forty-five minutes.

Well, closing time at the bar came at 2:00 a.m., and I was in no shape to drive home. The problem was, my friend forgot his pledge to me and was worse off than I was. I owned a van at the time, and I drove while my friend slept in the back. After about twenty minutes, I started to get the nods. I remember thinking, *If I can just close my eyes for two seconds, I'll be okay.*

....

"Michael, a battle is coming. Send angels to protect Brad and to keep his car from running off the road."

"Yes, Lord. I will bring others with me to intercept Brad's van on the interstate."

Just as the van was about to hit the guardrail, the atmosphere grew eerily silent.

"Here come some of Satan's warriors!" yelled Michael. "Get your swords ready. We cannot lose this battle!"

The sound of metal upon metal clashed in the heavenlies.

"Gabriel, you take the one with the bloodshot eyes, and I'll take the other two!" directed Michael.

Sparks were flying into the night as the celestial battle raged. Ether and the other two demons, however, were no match for Michael, Gabriel, and the other angels. Without any notice, the demons suddenly vanished as quickly as they had arrived.

"Gabriel, lift the van at a ninety-degree angle to the guardrail."

"That will look odd, Michael."

"I want it to look odd. I want Brad later in his life to remember this. I want him to know that the only way his van could have ended up there was because of us, Gabriel."

Later, in the presence of the Lord, Gabriel spoke. "Satan still wants Brad's life to be taken before it is time. How do you always know what is happening with everyone in the world?"

"I am everywhere, Gabriel. I AM the Lord. But people's prayers also bring more angels to fight against evil. Brad's aunt and uncle were praying for him tonight. It is not for you to understand, Gabriel, but prayer puts many things into action."

…………………

"Ether, what happened? Why are you back so early? You had a job to take the life of that young man tonight," demanded Satan of his minion.

"We arrived at his van, Prince, but there were too many angels around his car. We put up a fight, but there were too many of them," Ether attempted to explain.

"Ether, find out who is praying for Brad and then attack them. We don't want anyone to pray for him. But for those who insist on praying, keep them saying the same prayers over and over. There is no power in those prayers. If people knew what happens in the heavens when they call out to J for help, we would be instantly defeated. Keep people busy, Ether; then they will have no time to pray."

....

All of a sudden, I startled. I had fallen asleep at the wheel but was suddenly jolted awake. Directly in front of me was the guardrail. The van was in drive and its bumper touching the guardrail, but there were no scratches on any part of the vehicle. The only way my van could have gotten into that position was if I had backed it up and then driven directly into the guardrail, or if it had somehow been placed there by someone.

Fortunately, there was a large shoulder on the road, and the van was not hanging out onto the interstate. We were on a desolate part of the highway, and it was around 3:00 a.m. I backed into the eastbound lane and then drove home. Never once did I think about this instance till years later. It was, however, a wake-up call for me to watch my partying.

Terror

"Gabriel, we need to thwart Brad's plans in every direction. He needs to see that his only hope is in me. I confuse the ways of the wicked and the schemes of the wise so that they might turn to me."

"Then why doesn't every one turn to you, Lord?"

"Some hardened hearts will never change, Gabriel. Some face many trials in their lives that I have allowed but still trust in their own ways."

"Even if they are miserable, Lord?"

"Yes, even if they are miserable."

....

I seemed to be hitting a brick wall everywhere I turned. Never once did I think about why. I was just determined to keep my head down and keep plowing forward even if it was in the wrong direction.

....

"Panic!" called Satan to one of his demon cohorts.

"Yes, Prince."

"I need you and Ether to take Brad to the brink of ruin. He has always trusted in his own health for his security. I want him to think every day that what happened to his dad is going to happen to him. I don't want him hurt—I want him destroyed. I'm still not sure what J sees in Brad, but I am going to destroy him before J gets a chance to save him.

"Now use your talents to have Brad focus entirely on himself. Make him selfish by causing him to think only about how to survive. Consume his life with fear so he will be of no use to anyone else."

"Yes, Prince. Consider it done."

....

At this point in time, something troubling began occurring regularly in my life. Since the night of my initial reaction to marijuana, I had suffered from panic attacks. The first attack happened while I was driving to a ski resort in New York. I had to get out of the car to try to catch my breath. My heart was racing, and I was scared to death for no reason at all. At first these attacks came only once a month. A year later, they were happening numerous times every day. It eventually got to the point where I could hardly drive. I was now a prisoner of an evil adversary that I could not see.

For the next three years, my life was littered with failed relationships, poor grades in school, and a sense of urgency on one hand and despair on the other. I met a girl who had a young child, and shortly after we met, we married. I thought I loved her, but as I look back now, I realize I married her mainly because I thought I was going to die at a young age and wanted to experience what being married was like. That was one of the most selfish decisions I have ever made, but at the time, I didn't see it that way. I just thought I'd better try to enjoy my life now and find something to make me happy. In the back of my mind, I kept wishing someone would tell me what I was doing wrong to bring all this pain upon myself.

One night Andrea, my wife, wanted me to go to a bar with some of her friends to have my fortune told. I sarcastically thought, *Yeah, right,* but I was curious, so I agreed to go. In the back of a faintly lit section of the bar sat a woman at a card table. Her eyes were sunken, and she looked as if she had not slept in days. She first looked at my hands and then at some cards placed in front of her. Shivers flew up and down my spine when she said, "Your dad died at a young age. You miss flying, but you will get your flying career back. However, I see much pain and turmoil in your life."

As I will explain later, I had begun a career in flying but had been forced to relinquish it. Though the part about flying again was promising, the part about pain and turmoil was disturbing. I remember thinking, *Hey, I paid for this. Shouldn't I be getting some good news?* So I asked the woman the color of the uniform I would wear when I flew again. At the time, I was trying to get on with US Airways, whose uniforms were black with silver stripes on the sleeves, but she said the color of the uniform was blue with gold stripes. She also said many more things that night that only I would know. I came away from that encounter feeling as if I had met a relative of the devil. I was not saved at the time and knew nothing about demons. I also did not know that God allows people to see only so far into the future.

I went home and told Andrea about the woman's predictions and her description of the uniform. Since I was going though these horrible panic episodes and did not have much hope, my mind swirled in hyperspeed. My only conclusion about the uniform was that it must mean I would be a skycap at the airport and check people's bags, since the colors described were the colors of the skycaps' uniforms. Gloomy and in a poor frame of mind, I fretted, *Why did I ever go to that woman? What a stupid idea!*

After a few years of marriage, my wife began wanting to go out at night to try to unwind. Because of my panic attacks, I could hardly move, much less go out and socialize. To Andrea's frequent requests, I always answered, "I can't go out tonight. I feel like I'm going to die."

Then Andrea would say something like, "Brad, every night you feel like you're about to die. I can't take it anymore."

Eventually my wife decided to start going out alone, saying, "I need to go out, even if it is by myself." This was the wedge that eventually caused our marriage to fail.

My life was unraveling, and I couldn't understand why. Had I believed in God, I would have understood that someone with a sinister mission was trying to destroy me. But in the end, I always attributed my problems to having done something wrong that I was now paying for.

I was only twenty-three years old and at the lowest point of my life. I could barely leave my home, so crippling were the panic attacks. I actually cried out to God for understanding. I went to the Cleveland Clinic, and they ran a battery of tests; but they were all negative. I also went to countless doctors, but no one had an answer.

How could I have been such a disciplined runner yet be so incapable of controlling these attacks? They left me so frazzled and hopeless that I very nearly concluded that I would rather spend my days doped up on Valium in a mental institution than to put up with these panic attacks any longer.

Too sick to go to work one day, I turned on the TV at ten in the morning. A doctor on a talk show was talking about panic attacks. He described me perfectly. I called the network, got the doctor's number, and made an appointment to see him in a month. A ray of hope pierced my dark world.

When I went for my appointment, the doctor told me that I had an excess amount of adrenaline flowing through me at all the wrong times. My flight-and-fight response was damaged, he explained. He told me that he was going to prescribe a medicine to stop this excessive flow of adrenaline and that it would help remedy the situation. Within a few weeks, I noticed a dramatic change. Within a month, I was able to drive again, and within six months, my life was back to normal. I was back in the game, with no fear of anything out of the normal. I had my entire life back again.

With the dramatic turnaround in my life, I forgot about my prayer asking God for help and now felt like the master of my destiny.

Dreams Become Reality

Ever since I was a young boy, I had always wanted to fly. Hanging out around airplanes and airports was a dream come true for me. I now sent in an application and enrolled at a flight school in Oklahoma. Andrea, Chip (her son), and I moved out West. This was a chance for a fresh start.

After the first seven days of flying, my instructor and I were doing touch-and-gos, which are takeoffs and landings without stopping. My instructor asked me to come to a full stop, and then we taxied to another part of the airfield. As he opened his door, I was wondering what was going on. Then he announced quite matter of factly, "Brad, you are ready to solo. Do three touch-and-gos, and I will meet you back at OPS. (Flight Operations)

"Cessna 128CZ, taxi."

"Roger. 128CZ taxi into position and hold."

"Roger. Position and hold."

"Cessna 128CZ, cleared for takeoff. Make right traffic for runway 18."

"Roger. Cleared for takeoff; right traffic for 18."

Okay, no turning back now, I thought as I pushed the throttle forward. The plane accelerated and was airborne so quickly that it caught me off guard. The instructor had failed to mention that a small plane carrying two hundred pounds less than before would lift off more quickly.

As soon as the wheels left the ground, a sweet peace enveloped me. This was what I was meant to do. After three touch-and-gos, I landed and taxied back to OPS. The other students and instructors were waiting to initiate me into the flying ranks. Anytime someone completes their first solo flight, the back of their shirt is cut in half and signed by everyone.

In the months ahead, there was much to learn in flight and ground school. The day consisted of four hours of ground school in the morning and two to three hours of flight training in the afternoon. I had always received Cs and Ds in high school and college, but now I was making straight As. From this I realized that I wasn't as dumb as so many people had thought I was. Later I would learn that my double whammy of ADD and dyslexia was often found in entrepreneurs.

One of the most important parts of my training was learning how to get out of a stall. A stall happens when the airplane itself quits flying, not when the engine quits. Lift on an airplane wing is caused by higher pressure under the wing and lower pressure above the wing. Both these areas of pressure trying to equalize creates lift.

One way to reduce the speed of the airplane for a stall is to hold the same altitude and slowly reduce the throttle to idle. If the same altitude is maintained, the plane starts to shake, which is the beginning of a stall. When the stall is in full swing, the plane drops because there is no lift.

As students, we always practiced the stalling maneuver at three thousand feet and above to give ourselves plenty of time to recover if there was a problem. When the aircraft is low and slow, such as at landing and takeoff, there is no room for error. We thus practiced the maneuver hundreds of times at higher altitudes until it became second nature. Recognizing the signs of an impending stall before it actually happens is the key to preventing one.

In less than a year, I secured my private, commercial, instrument, multiengine, certified flight instructor, and instructor certificates. Then I was hired as a flight instructor, which meant I could gain flight time without having to pay for it.

As I soon learned, nothing in my life was ever normal, and that was evident in what happened with my first student, a young man from Belgium named Shawn. The plan was for me to take Shawn out in a Piper Arrow to get him qualified in a complex airplane. Shawn preflighted the airplane and told me everything was a go. After we were airborne, I directed him to the practice area where we could level out at 4,500 feet for maneuvers. Shawn did everything I asked him to do: executing left and right turns while holding the same altitude, accelerating and powering on stalls, and doing a mock emergency landing. He did everything with precision, and then I told him to head back to the airport.

As we approached for landing, I instructed Shawn to go through the prelanding checklist. While I was looking out the window and checking for air traffic, Shawn suddenly tapped me on the shoulder. I looked over and saw Shawn's hand on the gear lever, moving it up and down. However, even though the lever moved, the landing gear would not. The gear was locked in the up position.

At times like these, I would often talk to God, though I never expected him to listen. I would always say, "What next?" Well, this was a perfect "what next" moment.

I tried all sorts of maneuvers to get the gear down, but to no avail. Fortunately, the Piper Arrow has a great safety mechanism to force a gear stuck in the up position to drop down by the force of gravity. I tried that lever and heard a big thud, indicating that the gear had dropped.

The next thing any pilot looks for when the gear comes down is "three green." The nose wheel has a light, and there is also a light for the gear under each wing. When you call out "three green," this means all the gear is not only down but also locked. Well, on that day with Shawn, the gear came down, but we had no green. What next?

"Roger, Ardmore Tower. Piper 841RJ needs to declare an emergency. Our gear seems to be down, but we don't know if it is locked. Requesting a low pass to see if it is locked."

"Roger, Piper 841RG. We will look with our binoculars to see if the gear is down, but we won't be able to tell if it is locked. Fire trucks and all emergency personnel are in position."

"Shawn, you'd better let me take it from here," I told the young man.

The low flyby revealed the gear was down, but no one knew for sure if it was locked. This would have to be one of my smoothest landings ever. If the gear was not locked, any large vibration would collapse it, and then the crash would be on.

"Piper 841RJ cleared to land runway 18. All emergency vehicles in position."

"Roger, Ardmore Tower. 841 RJ cleared to land runway 18."

With a hundred feet to go, it was all business. The flare-out was smooth, but I didn't want to touch down till the airplane was at its slowest. The main wheels were now down—so far, so good—and as the speed slowed, I gingerly lowered the nose wheel. Yes, the wheels were locked, and we taxied back to the terminal.

"Thanks for all your help, Ardmore Tower."

"Roger. 841RJ, come up and see us after your debrief."

"Roger. Looking forward to it."

I really didn't learn to fly until I started flight instructing. By seeing other people's mistakes, I learned how to do things the right way.

I gradually gained flight hours, and my goal of becoming an airline pilot was in sight. I still needed another thousand hours of flight time, but an even greater problem loomed. Most airlines required their pilots to have 20/20 vision without glasses. The problem was, I wore glasses. I thus set up an appointment with an eye doctor, having heard that some contact lenses could reshape your eyes. After wearing the contacts for a year your eyesight could be corrected to 20/20 without the use of any contacts or glasses. Did I say I was gullible?

....

"Lord, Brad's experience as a flight instructor shows a lot of wisdom in how you are trying to teach him and others," remarked Gabriel.

"Man learns more from his failures than from his successes, Gabriel. Failure stings, and over time, failure is a catalyst that brings a man to repentance."

"How is that, Lord?" asked Gabriel.

"Man has always wanted the easy way in life and the easy way out of trials. But shortcuts never work, Gabriel. It is to a man's honor and advantage to spend his time learning and to then work hard with what he has learned. No man will ever be self-fulfilled until he works as unto me."

"But how do you teach that to man, Lord?"

"Trials and time, Gabriel," Jesus explained before continuing. "People who are very successful in a worldly sense are the most difficult people to soften. They think that they are the ones who gathered their wealth and acquired their good jobs. They never think for a moment that it is I, Gabriel, who controls their steps, their life, and even their next breath.

"Riches bring far more people to the depths of hell than to the gates of heaven. It is not the money itself that is the stumbling block; it is what it does to someone who isn't ready for it when it comes. Brad will have to learn this the hard way. I know his heart, and that is the only way he will learn—through many hardships and trials."

....

"Brad's training is going real well," commented Ether, "and many airlines are starting to hire."

"I know," answered Satan. "I have something in store for Brad that will sideline him for a while. He wants too much too soon. He is not willing to wait for anything.

"Man's lack of patience is one of the best tools I have to disrupt his life, Ether. He does not wait on what J has for him, and I am more than willing to give him something quicker that will fill that need. Just watch this one."

Are You Kidding Me?

Since the seventh grade, I had worn glasses. As I said, I wanted to fly, so I made an appointment with an eye doctor near Dallas, Texas, to see about getting contact lenses. After my examination, the doctor told me of a new procedure called radial keratotomy, or RK. This was 1978, and doctors were still experimenting on monkeys to refine the procedure. This doctor had just returned from training in Russia, and he told me that the surgery would allow me to see without either glasses or contacts.

Now as I said, I am a very gullible and impatient person, so when the doctor asked if I wanted the surgery, I jumped at the chance. Since this was a new procedure, the doctor didn't exactly have eager patients breaking down his door for it. So within only a couple of weeks, I had my surgery scheduled.

On the day of the surgery, Andrea drove me to the hospital. Some eyedrops were applied to numb my right eye, and the procedure began. In order for a nearsighted person to be able to see at distances, the cornea must be flattened. In my procedure, eight precision cuts around the cornea would reshape the eye to give me 20/20 vision without glasses.

All I could see was the scalpel coming toward my eye for each of the incisions. I could feel some pressure each time, and then the doctor would move to the next cut. The surgery lasted an hour, and Andrea drove me home afterwards. I wore a patch on my eye, but it was removed the next day. For the first time in ten years, I had 20/20 vision in my right eye without wearing glasses. The procedure for my left eye was scheduled for two weeks later.

While I was off flying status, I taught ground school. I couldn't wait to get the next eye done. I would do whatever it took to get on with the airlines, even undergo a risky operation.

The same procedure was planned for the left eye as was done for the right eye. But about twenty minutes into the operation, the room became very quiet, and the doctor's tone grew serious. "Brad, we've hit a soft spot in your cornea, and your eye is losing fluid. We will have to suture your eye and leave the stitches in for ten weeks."

Six times I had to endure a hooklike instrument piercing my eye and pulling the thread through for the suture. After they sewed me up and the Novocain wore off, the pain was unbearable. The suture was a very fine wire suture, but it still cut my eyelid so badly that I would just scream in pain. The only relief from the pain came when the eye eventually glazed over with infection.

Because the pain and sensitivity to light were so bad, I struggled to teach ground school. Most of my waking hours were spent in a dark room. Even with a patch over my eye, I found sunlight unbearable. Ten weeks passed at a snail's pace, and the doctor was ready to perform the operation again.

"Okay, Brad, I know you'll be glad to get this over with," said the doctor before he began. "First, we will do the cuts again, and then we'll take out the sutures."

The operation itself went great, but—and there is always a *but* in life when things are about to change—when the doctor began to remove the sutures, silence fell upon the room and people began scurrying about.

"Brad," explained the doctor, "that last cut put so much pressure on the eye that it opened up again." I swore and tried desperately to get off the table, but that was the last thing I remembered. I woke up the next morning in the hospital with sutures in my eye again *for another ten weeks.*

"God, why, why would you do this to me?" I bitterly complained. "Haven't I been through enough? What have I ever done to deserve this suffering?"

Now I would be off flying for a total of twenty weeks, with a chance of not being able to see again out of my left eye. Strangely enough, I was more worried about not flying than about losing the sight in my left eye. After my twenty weeks of sutures, pain, and infection, the doctor did the operation *again,* and all turned out well. Within two weeks, the pain of this bad memory was long gone and I was cleared for flight status.

....

"Lord, Brad cried out to you only because he was in pain, but at least he acknowledged you," Gabriel observed.

"Many people have a temporary faith in me, Gabriel. When they are sick, in a bad relationship, or in jeopardy of losing a job, they cry out to me for help. But when things turn around, they forget about me and go on with life in their own strength.

"Brad's faith will eventually turn from a temporary faith to a saving faith, but as you have seen, Gabriel, right now he has a hardened heart that is only interested in what he can do for himself. The time will come for change, Gabriel; just wait for it. But Brad's suffering must continue for a while until his heart is softened. When he finally realizes that I am in control and he is not, then and only then will he believe and be transformed."

....

"Prince, you haven't brought Brad's name up much lately."

"Have you been watching him, Ether and Panic? He is doing a great job of self-destruction without our help. He thinks he is controlling his destiny. He thinks that his will alone will make him succeed. As long as he goes down this path of self-centeredness, we can just watch him self-destruct. Our time is more valuable attacking those who are trying to pray to J."

Hired

I was flight instructing when the call came. I had been selected for an interview with Worldair Airlines. This was the opening I had been waiting for since I was a boy.

I sat outside the chief pilot's office as I waited for my interview. While I was waiting, out walked a couple of pilots from the office. All of a sudden, chills traveled down my spine. They were wearing blue uniforms with gold stripes. My mind jumped back to the fortune-teller. Maybe this was it, maybe this was a sign!

Fortunately, I passed the initial interview and was invited to a four-week training process. This process was an unpaid training program, and if it went well, I would be hired as a first officer.

The training was an awesome experience, and I learned all the new systems on the aircraft. The only time the planes were available to train with was late at night, because they were used on line during the day. I passed my oral, written, and flight exams. I then was given my wings and fitted for my uniform.

Preflighting the aircraft on my first flight was a dream come true. There I was, out on the tarmac at a major airport getting ready to fly for a major airline. The chief pilot needed to check me out in real flying situations with passengers on board, a process called a "line check."

“Okay, Brad, I have the first leg to Detroit tonight, so you get the radio. First flight in the morning, you’ll fly and I’ll work the radio,” the chief pilot directed.

I tuned into ATIS: “Cincinnati weather, 10,000 scattered, 5,000 broken winds light and variable, temperature 81, dew point 65, altimeter 30.05, using runway 27 left. Advise you have Romeo.”

“Cincinnati Ground, Worldair 1521 taxi with Romeo.”

“Roger. Worldair 1521 taxiing to runway 27 right via Juliet to Kilo; hold short of 27 right.”

“Roger. Juliet to Kilo, hold short of 27 right.”

I discovered that the hardest part of flying was finding your way to the airport terminal or the active runway once you landed. Each airport had a detailed map that outlined all the taxiways, runways, and access roads. But on a snowy night, you had better know what you are doing.

“Cincinnati Ground, Worldair 1521 holding short of 27 right.”

“Roger, Worldair 1521. Cross over runway 27 right; hold short of 27 left. Contact Tower on 118.3.”

“Roger. Cross 27 right and hold short of 27 left, Worldair 1521.”

“Cincinnati Tower, Worldair 1521 holding short of runway 27 left.”

“Roger, Worldair 1521; taxi into position and hold.”

“Position and hold, Worldair 1521.”

“Worldair cleared for takeoff. Climb to 1,800 feet; turn right to 360 for the Dayton departure route.”

“Roger. Cleared for takeoff, Worldair 1521.”

“Brad, help me bring the throttles to max take-off speed,” the chief pilot instructed.

“Roger. 40 knots, 70 knots, 100 knots V1, 120 knots rotate, V2, positive rate, gear up.”

“Roger, gear up. Climb checklist. Flaps up, autofeather off, check; power reduced to climb power, check; fuel balanced, check.”

"Worldair 1521, Cincinnati Tower contact Cincinnati departure on 124.95 cleared to 4,000 feet."

"Roger. Departure 124.95 cleared to 4,000 feet, Worldair 1521. Cincinnati departure Worldair 1521 heading 360 and climbing to 4,000."

"Roger, Worldview. Traffic at 11 o'clock, a Delta 727. Advise you have him in sight."

"Roger. In sight, 1521."

"Worldview, climb and maintain 9,000 feet, heading 360."

"Roger. Leaving 3,500 for 9,000 feet, Worldview 1521."

"Okay, Brad, what do you want to do tonight? There is a great bar at the hotel," suggested Bill, the pilot.

"Well, I think you just planned our evening, but it sounds good to me. We should get in just in time for the action."

"Worldair 1521, contact Cleveland Center 124.85."

"Roger. Cleveland Center 124.85, Worldview 1521."

"Brad, someday I want to get into a Lear jet and go to the right instead of the left," remarked my partner.

"What do you mean, Bill?" I asked, puzzled by his words.

"I want to have enough money so that I can sit in the back in my own jet and have someone else fly me," he explained.

"So is that success, Bill?" I asked.

"Isn't it to you, Brad?" he countered.

"Success to me, Bill, is being able to fly for the rest of my life. Maybe success for you would be to have a jet and have someone else fly you, but I could never give up flying. It is my life."

"You're weird. You know that, Brad?"

"Flying is it, Bill," I emphatically responded. "I can't picture myself doing anything else—ever!"

After about forty-five minutes of talking about anything and everything, I asked Bill, "How did you get to be chief pilot?"

"Being at the right place at the right time and playing the politics game," he answered. "How about you, Brad? How did you get started flying?"

"It's a long story, Bill, but we have some time over the next four days, so I'll fill you in."

"Worldview, turn left heading three three zero; intercept the Motor City arrival. Advise you have hotel."

"Just a second, Bill. Let me get this," I said and turned my attention to the task at hand.

"Detroit weather 1900 Zulu, temperature 79, dew point 80, winds 050 at 10 gusting to 20, landing runway 3 right, altimeter 29.94. Advise you have hotel."

Getting back to our personal conversation, I asked, "Are you married, Bill?"

"Yes, I've got a great wife and three kids. How about you, Brad?"

"What time are we going to get to the bar, do you think?"

"Boy, you switched that subject quickly," Bill laughed. "I know a story for another day when I hear it."

Worldair 1521, contact Detroit Approach on 119.45; descend and maintain 4,000."

"Roger, Cleveland Center. Leaving 9 for 4 and contacting Detroit Approach on 119.45. Detroit Approach, Worldair 1521 descending to 4,000 feet; have hotel."

Dive . . . holy . . . unbelievable!

"Worldair 1521, traffic 12 o'clock, one mile!"

"Yeah, thanks, Approach. The guy was wearing a red tie, if you would like to know," I sarcastically answered.

"Sorry, Worldair 1521. He must have popped up from a small airstrip in the area."

"Roger, Approach."

After our hearts settled down in our chests, Bill and I didn't say another word, except for the essentials. The small two-seater aircraft had passed within fifty feet of our windshield. It was here one moment and gone just as quickly the next. On the way to the hotel in the airport van, I remarked to Bill, "Well, maybe we haven't been so bad after all. That was a close call. Someone is watching out for us."

....

"Ether, I thought you had Brad's plane. That other pilot was right on course to take everyone out."

"I know, Prince, but at the last second, angels with swords distracted us, and we lost our focus," Ether explained. Then with a puzzled look on his face, he addressed his superior: "Why does J want anything to do with Brad anyway? His life-style certainly has no value for J."

"Something must be up for J to send his angels like he did, Ether. We better not miss next time. We don't get many chances like this, so we've got to make the next one count."

"Yes, Prince, I won't let you down."

....

The four-day trip was uneventful. I got to know Bill better, and we had some great conversations about life. I also got my feet wet as a first officer. Although I felt God had taken both my dad and my running, I thought I was finally in control of my own destiny.

Flying enabled me to forget the pain of the real world. When I was up in the air, nothing on the ground mattered. But of course, the problems were still there when I returned.

When I arrived home after my first trip, things at home were not going well. My marriage was rapidly deteriorating. It was all I could do to hold my own life together, much less take care of a wife and stepchild. Selfishness on my part was tearing our family apart. My wife did not understand my internal struggles, and I eventually just quit trying to make her understand my inner pain. Something needed to happen—and quick.

"I can't take this anymore," said Andrea. "For the past three years, you never want to do anything but sit at home. I like the job I have, and you like the job you have. We have been on different paths for a while. I think we need to finally put an end to this charade."

With that Andrea and I decided to call it quits. After eight weeks away from each other, we knew that we were right: the marriage would never work. It was an amicable divorce, since in reality the marriage had been over for quite some time. But the failure of my marriage created doubts in my mind whether I could succeed at any relationship.

....

"I hate divorce, Gabriel," said Jesus. "One of the main causes of divorce is self-centeredness. Man wants what he wants when he wants it. Brad is going to have to learn the hard way, Gabriel, that I am in control and that he is not the master of his destiny."

"Why does he lose every relationship he is ever in, Jesus?" asked the puzzled angel.

"Brad doesn't realize it, but he refuses to let anyone get close to him, like he did with his father. That way when they leave, he won't be hurt again. He never wants to feel that kind of pain again, so when women get too close to him, he shuts down."

"So does that make it right, Jesus, for him to cause others pain instead?"

"No, it doesn't, Gabriel, and Brad, through much suffering of his own, will learn this over time. But in this world, no one can truly love another unless my Spirit lives inside of them. Everyone in the world, Gabriel, has a soul, but not everyone has my Spirit. Brad is a long way from calling out to me for help, asking for forgiveness of sins and asking me into his life. That is the only way to be saved, and once a person is saved, I send my Holy Spirit to dwell inside of them forever."

"Is Brad one of the elect, Jesus?"

"All in good time, Gabriel, all in good time."

....

Flight 1992, Continued

"Full power, Leo, full power! It feels like this plane wants to stall!"

"Brad, we are way above stall speed. I know that, so just trust me," he soothed.

"No more flaps and keep the power up, Leo."

"Okay, Brad, you've got three miles to the end of the runway. Power back."

"No, there it goes again, Leo. I'm telling you, this plane wants to stall!"

"Okay, Brad, your call."

"Indianapolis Tower, Worldair 1992 cleared to land runway 5 right; winds 050 at 10, gusting to 18 knots."

"Roger, Indianapolis Tower."

"Cleared to land runway 5 right, Worldair 1992."

"Runway in sight, Brad; you can come off instruments," said Leo.

"Roger, Leo, runway in sight. This is going to be a fast landing, and the runway is snow covered. Let's pray these brakes work. I am not a praying man, but tonight I may try it."

Thankfully, we made a smooth landing, and braking was good on the snow-covered runway.

"Worldair 1992, taxi via Lima to Echo; contact Indianapolis Ground on 123.4."

"Roger. Lima to Echo; Ground on 123.4."

"Man, what a ride!" I exclaimed, turning to my partner. "I can't wait to get out and see what was going on."

"Me too," replied Leo. "Either you're a really good pilot, or I'm going to have to question your judgment."

As the ground crew directed us to the gate, the flagman seemed to focus on the right wing and started to call other people around.

"I wonder what they're looking at, Leo," I mused.

"I don't know. Let's get these passengers off and find out for ourselves."

After the passengers deplaned, Leo and I walked around to the other side of the airplane and couldn't believe what we saw. When ice forms on an aircraft, it usually forms as a thick covering over the wings. On the wings of our plane, however, the ice had formed vertically in ragged icicles, each approximately five inches in length. There were about seven of these icicles on each wing.

"Man, that is evil looking, Leo. I don't know how we got it down," I remarked.

"Well," Leo began, "I know you won't believe it, Brad, but it was by divine intervention."

Ignoring his remark, I sniffed the air and said, "Hey, Leo, do you smell that?"

"What, Brad?"

"Smells like ether." I had been given ether to put me asleep when I was five and needed some teeth extracted, and I had never forgotten that smell. "Wonder why I got a whiff of it now?"

"I've never seen anything like this in all my years of flying," marveled Leo.

"Leo, I think we deserve to get hammered in the bar tonight after this one."

"For once I agree with you, Brad. Let's get out of here."

The next few months were uneventful. Then one night around ten o'clock, something happened as Captain Tom Harris and I were flying from Nashville to Louisville.

"Louisville Approach, Worldair 1271; 7,000 feet with Echo."

"Roger, Worldair. Maintain present heading; descend and maintain 5,000 feet."

"Roger, Approach. Worldair out of 7 for 5."

Pow! The loud sound came out of nowhere.

"What the hell was that, Brad?" a startled Tom asked.

"Sounded like something hit our right wing, but what's up this high?" I answered.

"Feels like the ailerons are bound up, not free," Tom observed.

I turned the control column to the right and the left and felt the same thing. However, this fact alone would not affect our landing.

"Let's just leave the flaps at thirty degrees instead of full flaps, Brad."

"Okay, Roger that, Tom. Let's get on the ground and see what's going on."

As Tom and I taxied in, the agent directing the flight was looking intently at the right wing. *Well, he must see something already. This should be interesting,* I thought.

After the passengers deplaned, Tom and I walked around to the right wing. By now a number of other pilots and ground personnel were walking over too. To our amazement, there was a hole in the leading edge where the landing light used to be, and there was blood all over the wing.

"Man, I didn't know birds flew that high. Maintenance will probably tell us what kind of bird it was when they take this bird apart," I commented to Tom.

"What bird—our bird or the bird that is now just feathers?" Tom wryly responded.

"Okay, Tom, I think this deserves a trip to the bar. If that bird had hit our windshield, one of our heads would have been taken off."

"Yeah, that was a close one, Brad. Drinks it is, and I'm buying."

"See you in the bar," I promised.

The next day Tom and I got word that we had hit a mallard duck. It had broken through the landing light, and its insides had wrapped around the aileron cables in the back of the wing.

"Does this stuff happen to you all the time?" Tom asked me, shaking his head in disbelief.

"Tom, if you only knew!" I answered, shaking my head in response.

After twelve months of flying in all sorts of weather and devoting countless hours to my job, my number came up for captain. Now my life would be complete—or so I thought.

Being a captain for an airline meant I had reached not only the pinnacle of success in my profession but also the pinnacle of responsibility. I spent the next four weeks in an intensive process of ground and flight training. Ground school covered all the aircraft systems and instruments; pneumatic, hydraulic, and electrical knowledge; and all the specifications of weights and balances. It was essential for pilots to know this information frontward and backward and top to bottom so that it would become second nature. If an emergency occurred, a pilot had to act by instinct, as there might not be time for anything else.

The day for my check ride came, and I was as nervous as when I had run my first cross-country meet. Ground school went well that day, and then it was out to the aircraft to go through normal and emergency procedures. I took off, and the chief pilot handled the radio. All three thousand hours of prior training now came down to this one-hour flight.

After having me fly on one engine and partial instruments, the chief pilot directed me back to the airport. I didn't know if I had passed or flunked. The ground crew at the terminal directed us in, and I shut down the engines. The chief pilot looked at me, thrust out his hand, and said, "Congratulations, Captain. Come upstairs and let's get the paperwork done."

The pain of losing my dad, my running career, and my marriage somehow melted away in that moment. Though I had lost much, no one could take my flying away. This is what I was destined to do.

I was now ready to fly the line, but before beginning in my new position, I had four days off. Since I had free passes to fly, I decided to go to Las Vegas. It was party time.

Now What?

On the flight to Las Vegas, an odd, frightening feeling descended upon me, almost like an ominous storm cloud. But where was this off feeling of anxiety coming from? I finally had it all: I had attained my childhood dream of becoming an airline captain, I was single, and I could drink till dawn with no responsibilities to tie me down. But now I suddenly felt more alone than anyone else in the world. *How can this be? I have everything yet feel so empty. Hopefully, when I get out to Vegas, I can shake this feeling and unwind.*

Since my divorce, most of my nights when I wasn't flying were spent in a bar. I didn't care who I slept with or how much I drank. I felt that I deserved this out of life. Who was I hurting anyway? I rationalized.

Arriving in Vegas, I checked into the hotel and hailed a taxi to take me to the nearest brothel. Sorry to say, but I preferred to pay for sex than have to date someone for it. My feeling was that I would have to pay for it one way or another. My lifestyle was a train wreck coming, and the train wreck was not far away.

During my time in Vegas, I visited the brothel a couple of times and spent all night at the gambling tables, but still I felt lost. I boarded the next flight back to Cincinnati and shrugged off the bad trip as just a "feeling." Trying to put it out of my mind, I reminded myself that I was flying as a captain now. I was finally the captain of my destiny. No one could take this from me.

....

"Sexual immorality is my specialty," gloated Satan to Ether. "I have Brad so tied up in lust and addictive behavior that he is no good to anyone. He is so far from J that we don't have to worry about him. Just keep sending women his way and he will never be able to live a normal lifestyle. The scars will be too deep for him to ever forget his past."

....

Gabriel asked, "How can you stand to even look upon Brad's lifestyle, Jesus?"

"I hate sin," the Savior said, "but Gabriel, every man and woman is a sinner and saved only by my grace. Many people in the world feel the same way Brad does right now.

"I enable people to have riches and status, but it is only I, Gabriel, who enables man to enjoy these things. No matter how hard someone tries to enjoy the things of life, I will frustrate them at every turn until they know and believe that I am God."

"So this is just another piece of the puzzle for Brad to come to the end of himself?"

"Yes, Gabriel. I am slowly stripping away everything Brad has put in his life in place of me. That day will come, Gabriel, when his heart will change."

....

Flying the Line

"Toledo Ground, Worldair 1520 taxi; we have Alpha."

"Roger, Worldair 1520. Taxi Bravo; hold short of runway 7. Contact Toledo Tower on 118.1."

"Roger, Ground. Taxi Bravo, hold short runway 7, Tower 118.1."

Then, addressing the passengers, I said, "Ladies and gentlemen, this is your captain speaking. We are number two for departure. Weather in Cincinnati is overcast and fifty-two degrees. Flight attendants, prepare for departure."

"Tom, let's finish the checklist. Prop sync off, check; fuel pumps on, check; hydraulic boost pump on, check; anti-ice on, check; V1 will be 105, Vr 120, check; preflight check complete."

"Toledo Tower, Worldair 1520 holding short of runway 7."

"Roger, Worldair 1520. Taxi into position and hold."

"Roger, Tower; 1520 position and hold."

"Worldair 1520 cleared for takeoff. Turn right heading one eight zero; contact departure on 125.8."

"Roger, Tower. 1520 cleared for takeoff. Right heading one eight zero; departure 125.8."

Tom called out the airspeed: "Brad, 60 knots."

"Okay, control of rudder, V1, VR, rotate, positive rate, gear up."

"Brad, we have a red light for the landing gear. Toledo Tower, Worldview 1520 has a malfunction on the gear, so we will need to return to the airport. Toledo Tower, we need to do a low pass so you can assess any damage."

"Roger, Worldair 1520. Right heading one eight zero for vectors back to runway 7."

"Tom," I directed, "do the emergency checklist and inform the passengers."

"Got it, Brad," he quickly responded. Then he turned his attention to the passengers: "Ladies and gentlemen, we have to return to the airport. We have a warning light that we need to address concerning our landing gear. We need to make a pass by the tower for them to see if our gear is down. We will do a low pass, and then we will land."

"Tom, prepare the passengers for an emergency landing. Just keep the passengers calm."

"Got it, Brad."

"Worldair 1520, the gear looks down, but we will have emergency vehicles standing by."

"Ladies and gentlemen, we are cleared to land," Tom informed the passengers. "The gear is down, and we will be on the ground shortly. Assume the emergency position."

"Vref 120, got it, Tom. Will try to make this as smooth as possible."

"Worldair 1520, Toledo Tower cleared to land runway 7. All traffic at the airport is stopped, and emergency vehicles are in place."

"Roger, Toledo Tower. Cleared to land runway 7, Worldair 1520."

"Okay, Brad," Tom coached, "130, 125, 120, over the fence. Let the plane float a little longer, and just ease the main gear onto the runway. After the airspeed bleeds off a little, then gently put the nose wheel down."

I followed his instructions and breathed a sigh of relief as we touched down.

"Okay, great landing," Tom congratulated me. "Let's see what the problem is."

"Worldair 1520, contact Ground 121.9."

"Thanks for your help, Tower. Worldair 1520 to Ground 121.9."

"Toledo Ground to Worldair 1520, taxi to the gate."

"Roger. Worldair taxi Bravo to Alpha."

"Roger. 1520 Bravo to Alpha."

"Let's get the passengers off and see what's going on," I told Tom. "Call Cincinnati and see if they can get some maintenance guys up here."

"Got it, Brad. Let's get out and look under the wing."

"Long trip, guys?" joked Jeff, who had been directing company flights for over five years. He was always joking, but this time he quickly turned serious. "Hey, guys, you need to see something," he said.

Tom and I walked under the wing, and there it was. When the gear had begun to cycle, one of the rods that pulled it up had broken. The rod had punctured a hole in the wing and missed the fuel tank by only inches.

"You guys must be living right. Well, maybe I should say Tom must be living right. Maybe I should start too. Brad, are you always this lucky?" Jeff asked.

"I don't know if I'm lucky or not, but odd things do seem to always happen to me," I remarked. Then turning to Tom, I said, "Well we are going to have to deadhead back to Cincinnati. Our day is done. Let's get some coffee—and hey, does the beverage cart carry Baileys?"

....

"It's time to take Brad's flying away, Panic. He is finally starting to feel comfortable in the lifestyle he's living. We don't want him to get too comfortable and to start thinking about J. We've got to keep him on the run. It's time to stop this nonsense and take away his hopes and dreams. Can you do this, Panic, or do I need to send in someone else?"

"I won't fail you, Prince."

....

One morning I woke up early after an overnight in Cleveland, Ohio. I showered, put on my uniform, and went to the bathroom. It was then that I noticed that my urine had blood in it. I had suffered from a kidney stone a number of years prior to this, so I didn't want to take a chance and fly in my current condition. I called Flight Ops and told them I was going to the emergency room in Cleveland and that they needed to find another captain to take my flight.

The hospital did some initial tests and then referred me to a urologist. The urologist asked if I was on any medication. I told him that I was taking something for an adrenal-system imbalance. I had spoken to my personal doctor on numerous occasions about my flying and use of the medicine. Since I was running five miles a day, in great shape, and had no adverse reactions to the medication, he thought everything was fine.

The urologist then performed a cystoscopy on me—which, incidentally, hurts beyond anything you can imagine. I waited for the results for the longest time, all the while expecting the worst. Finally the doctor returned and said, "Captain Henry, you're bleeding a little from your right kidney, but that's not what concerns me. However, the medication you're on does concern me. I just called your airline and told them you were on drugs. You will never fly an airplane again."

"Are you kidding me, Doc?" I asked incredulously. "How about patient-doctor confidentiality?"

"I'm just trying to protect the flying public, Captain," explained the doctor.

Stunned, I left the doctor's office, sat down on a park bench, and cried as hard as I did on the day my dad died. I called the chief pilot, who was also a friend, and explained the whole situation. Instead of forcing the FAA to get involved, I turned in my wings and that was the end of it—no more flying for me ever again.

God hates me, I thought in despair. *He's taken away everything I ever liked or loved. What kind of a God does this?*

I haven't killed anyone or robbed a bank. Why this persecution? Why doesn't he leave me alone?

....

"Jesus, that was a tough one," said Gabriel. "Now Brad really thinks you're behind everything bad that ever happens to him."

"I am going to love Brad no matter what, Gabriel. I created him, and I know every cell in his body. Most importantly, I know his heart, just like I know everyone else's heart. No one can hide from me, Gabriel, no matter how hard or fast they run. Man always judges others by their outward appearance, but fortunately, I see the heart. The heart can be changed, Gabriel, but it is only I who can change it. Everything in the world I work out for *my* perfect purpose, for my good and the good of mankind."

"Brad wouldn't listen to you now, would he?" the angel sadly remarked.

"No he wouldn't, Gabriel, but one day he will."

....

"Ether, we can just sit back on this one. J doesn't know what he is doing with Brad. He thinks that all these trials will bring Brad to some kind of spiritual repentance. We are smarter than J, Ether. Brad thinks J has taken from him everything that matters most. Now we just need to move in for the kill.

"Another way we have snared Brad, Ether, is through his immoral lifestyle. We have ensnared him so he needs to look at pornography, needs prostitutes and women, and needs to get drunk. Brad is well on his way to hell and too far gone for J to save him. Another one down, Ether!"

"Yes, it certainly looks like it, Prince. Okay, how many more can we snare today through pornography and a lustful heart?"

A New Normal

I made some phone calls and started working for my uncle who was a chiropractor. He had made a back cushion that he wanted me to try to sell to other chiropractors. Though I did my best, the volume sold through targeting chiropractors just didn't bring in enough money for me to make a living.

On one of my outings, a truck stop bought a couple of dozen cushions from me. Within a couple of weeks, I received a call from the truck stop informing me they were sold out. I told my uncle that this was the route I wanted to take: selling back cushions to truck stops.

So off I went to every truck stop in the country. I would leave for two months at a time, come home for a week, and then leave again with a fresh supply of back cushions. After a year, we had secured most of the major truck stops in the country.

I had not studied marketing in college, and even if I had, I would not have remembered it. But it came to my mind one day to approach Target and Sears stores about selling our cushions. I thought it would be best to go to one location that had many stores than to drive to every store. The strategy worked, and after many appointments in Chicago and Minneapolis, we successfully secured Target and Sears to carry the cushions.

Now I had a job with financial security. God had taken away my dad, my running, my marriage, and my flying, but building a business was firmly in my control.

....

“Jesus, I would have thought that all the trials you have allowed into Brad’s life would have changed him by now. But he is just as zealous as ever for recognition and determined to honor himself through building his business,” observed Gabriel.

“It may seem to Brad like I am taking things from him, but I am building in him a faith that will one day be unshakable. A man doesn’t learn faith and trust through an easy lifestyle, Gabriel. The longer the trial in a person’s life, the more dependent he becomes upon me. Brad is stubborn, though, and it will take many long trials to bring him to his knees.

“I don’t want to speak out of line, Jesus, but that sounds harsh.”

“What is worse, Gabriel: for Brad to spend eternity in hell or for me to bring him to repentance here on earth through tough trials?”

“Yes, I see,” said Gabriel quietly.

“Brad will quit asking why one day and finally say yes to me.”

....

I began having some internal conflict with some of my family members in the business and knew after one heated meeting that things would never change. I met with my uncle, and we agreed that I needed to leave. We both wept, as my uncle had been like a father to me. I took the easy way out and left in the middle of the night to start a new business in Pennsylvania, the same type of business my uncle had. This decision pitted many people in my family against me, except for my uncle. He understood, but no one else ever did until many years later.

I patented an item called the Beach Buddy. This was a neck cushion for use at the beach, featuring a pocket for holding change and keys. My business associates and I offered it in neon colors, and it sold well. A large drugstore chain ordered forty thousand units, and we had to have them ready in six weeks. At the time, the Association for Retarded Citizens (ARC) and the Association for the Blind helped us with the manufacture, sewing, and stuffing of the pillows. In the last week of production, we were behind on meeting our deadline, so I stopped by the ARC to check on the progress. One of the tables was quite in disarray. Apparently, some of the students had been overmedicated and were using the pillows to sleep on.

The more I worked with the people at the ARC, the more I realized that wealth does not secure happiness. Here were people who could think only about today, this hour, or this minute, yet they had more joy than most "normal" people. That made me start doubting what "normal" really was. I felt so sorry for the parents of these handicapped kids, but to my surprise, most of them were joyful too, just like their children. I just couldn't understand this. Little did I know that years later I would have a handicapped child of my own. Was God getting me ready now for what was to come later?

The order went out on time, and I was finally happy—for a day. But from my exposure to the people at the ARC, I started to want what they had: happiness all the time. I began to realize that their happiness was not based on their circumstances, but mine certainly was. Was that the key? The answer would elude me for many years to come.

In the next ten years, I started three new businesses, all which barely broke even. For two of those years, I lived in my office since I couldn't afford to pay rent plus start a business. The *why*s of this life were growing more frequent, but I didn't know who to ask about them. I would look up to the sky and see the contrails of a plane and just ask myself, *Why, why, why? Why was that taken from me?* I had no answer for my tormenting questions.

I worked as hard as I could during the day and spent most of my evenings in a bar, trying to forget how much I owed investors and how many times I had failed in business and relationships. I tried to ignore the gnawing feeling that nothing could give me peace. *There has to be something more to life, or I might as well end it now.*

Panama City

I thought I might as well start running again to try to get back into shape. I had done some small-distance triathlons, which included a one-mile swim, a twenty-five-mile bike ride, and a six-mile run. But my ultimate goal was to do a full Ironman competition. An Ironman includes a 2.5-mile swim, a 112-mile bike ride, and a 26.2-mile run. Now I wasn't so stupid as to try doing a full just yet. Instead, I signed up for a Half Ironman in Panama City, Florida. I had six months to train for it.

The Panama City event consisted of an ocean swim of 1.2 miles, a bike ride of 56 miles, and a run of 13.1 miles. I trained enough so that I could do all the distances by themselves, but in the event, I would have to be able to put them all together. In preparation for the swimming part of the competition, I practiced only in a pool. *How hard can an ocean swim be?* I thought to myself. Did I say I was naïve?

On the morning of the swim, I awoke to what sounded like thunder. Actually, I only wished it were thunder; a cold front had come in during the night, and what I heard were the waves crashing onto the beach. A sick feeling in the pit of my stomach warned me of what I was getting myself into.

I put my wetsuit on and completed all my prerace duties. Soon it was time to begin. The competitors were sent out in groups based on age. My group had about 150 people in it. The swim was approximately nine hundred yards out, two hundred yards across, and nine hundred yards back to shore.

We started by a large pier, so I thought I could use the pier to help me navigate.

The gun went off, and there was no time to worry anymore. Large orange buoys marked the course every hundred yards straight out. But once I was in the water, I soon realized that the only time I could see the markers was when I was on top of a swell.

"Are you the official for this race", asked a pro athlete.

"Yes", said the official, what's the problem?"

"There's a strong riptide out there, and if we pro athletes can't swim in it, then I'm sure the less experienced ones will not be able to handle it. Look, it's even pulling the buoys out of place."

"That's what we're hearing from the scuba divers. We won't count the swim portion of the race. The problem is, we can't call the age-group athletes back in."

"Hey, that's your problem. I just hope everyone makes it back in safely."

I finally made it out to the farthest buoy and then swam diagonally to the last buoy. On my way back to shore, after swimming for about five minutes, I looked up and noticed there was no one around me. The last time I had looked, there had been at least twenty swimmers around me. I was so far out now, though, that the large hotels on shore looked like pin dots. The pier was far off to my left, instead of in front of me where it should have been. I thought to myself, *Well, maybe there are two piers.*

I continued swimming, but the more I swam, the more I felt like I wasn't making any progress. Even though I was wearing a wetsuit, the tide was so strong that it constantly pulled at my legs. I knew I could easily panic, but I also knew that I wouldn't make it back to shore if that happened. Finally I spotted the shoreline, and after another five minutes of swimming, I made it to dry land. Instead of being in the water for the expected thirty-five minutes, I had been in for almost fifty-six minutes. I later heard that not only had they canceled the swim, but they had also pulled more than twenty-five people to safety that day.

When I emerged from the water, I was so far off course I didn’t know whether to go left or right to retrieve my bike. Actually, I was so happy to be on shore I could have stopped the race right there. Eventually I figured out which way to go and located my bike among the handful of bikes left. I began the bike course just thankful to be alive. I ate and drank at all the aid stations and was feeling pretty good. The problem was, I was feeling too good. I didn’t pace myself and pushed too hard. After completing the fifty-six miles, I dismounted my bike, exhausted. It was not the best way to start the run.

At mile five of the run, I began to see double. I saw a guy watering his lawn, and I just sat down on his grass. He walked over to me, and I said to him, “I can’t believe I paid sixty dollars to do this.” I got up and ran another mile, but then I called it quits. The safety vehicle picked me up and dropped me off at my hotel. I was so dehydrated that I couldn’t make it home and spent two days in the hotel, nursing a massive migraine.

This race taught me a lot about life. No matter how much we prepare for something, there are always things beyond our control. Actually, my life was a lot like that. I was trying to be in control of all things, but I would soon realize that I was not in control of anything. My life was spiraling out of control, and this race just pushed home that fact. Little did I know that my fight over control would come to a head five months later.

Heaven's Destiny

One night as I sat alone in an apartment building, I had taken all I could take. My business ventures had failed, along with my relationships and everything else I had touched. *What's the use?* I thought in despair. Though I had considered taking my life numerous times before, I didn't want to hurt my brother or mother. Additionally, I realized that because of my self-absorption and selfishness, I had left a trail of broken relationships behind me, and I didn't want to leave this earth in that way. Sitting in my apartment that evening, I finally yelled at the top of my lungs, "Why?" but there was nothing.

I turned on the TV with the intention of watching the comedy channel. I clicked through all the channels and then went through them again as I looked for the station. Finally I found a program where a guy was flailing his arms and walking as if he were drunk, so I thought I had found the comedy channel. As I listened to this man, however, I was mesmerized—but not by a comedic performance.

The man speaking was David Ring, an evangelist. I was thirty-eight years old and had *never* heard the gospel of Jesus Christ. David Ring was a man who had suffered tremendously with the loss of his father and mother at a young age. On top of that, he had cerebral palsy. As I watched the program, he kept repeating, "I have cerebral palsy. What is your problem?"

The evangelist's words were like a knife cutting open my heart and laying it bare for me to see. That night, November

29, 1992, was the night God had ordained for me to hear the good news of salvation. That night, at 7:30 p.m. and with great rejoicing and tears, I became a child of God. I asked for forgiveness for all the many sins I had committed. I acknowledged who Jesus was and what he had done for me on the cross, and then I asked him to come into my life. In that moment I, Brad Henry—the failure, the drinker, the sinner—became the child of the King.

....

"Ether, what the . . . ? How did you let this happen? Didn't you see it coming? The fire of hell won't be punishment enough for you if we lose one we had so firmly in our grasp. This is unacceptable, and you will pay for this with your life!" ranted Satan.

"There is only one way you can redeem yourself in front of me, Ether. Brad is saved for eternity in heaven, which makes my blood boil. We cannot ever take away his salvation, but I want it to be your life's mission to never allow him a moment's joy. I want him persecuted from morning to night. I want his depression to increase, and I don't want him to take anyone to heaven with him. Do you hear me, Ether?"

"Yes, Prince," the demon meekly responded, "I do. Consider it done."

"It better be," Satan snarled. "You know your mission is to create havoc and discontent. I have also asked Confusion and Panic to help you thwart any plans Brad may have of getting to know J better."

....

"Jesus, what an hour it has been! Angels have been rejoicing over Brad's decision to accept your free gift of salvation. Your plan is always perfect, Jesus, but I thought Brad was too far gone for you to save him," admitted Gabriel.

"That is what Satan wants everyone to believe, Gabriel. He wants people to believe that they are too bad, have sinned too long, and have hurt too many people to ever ask for forgiveness. But in Brad's case, he didn't have anywhere else to go."

"So is that it, Jesus? Nowhere else to go?"

"That's part of it, Gabriel. That's why prosperity chokes the life out of people. Prosperity, for the most part, replaces the desire for me. Nevertheless, I have spared Brad's life for this day. He will follow me and testify to my saving power and cleansing grace. And when he finally sees his dad in heaven, it will seem like only a day since they last saw each other. It is the same for any believer in me, Gabriel. Remember, I am not bound by time, and neither is the one who loses his life in me."

....

"Confusion and Panic!"

"Yes, Prince."

"Don't let Brad go into that Christian bookstore. I don't want him to grow in this new faith. I want him to remain ineffective. If he doesn't read a Bible, then he will not have power. But just as important, make sure he doesn't learn how to pray because that will release even more power. I already have enough angels to fight; I certainly don't need any more. Prayer unleashes more angels anytime a believer prays, Confusion, so it is your job to keep Brad so busy chasing after business ventures and women that he has no time to pray. Got it, Confusion?"

"Yes, Prince. Consider it done."

....

I wanted to purchase a Bible and read about God. Man, what was happening to me? I finally summoned the courage to go into a Christian bookstore to obtain some materials. To tell you the truth, at this stage in my life, I felt more comfortable in an adult bookstore than in a Christian one. As I walked into the Christian bookstore, I felt that everyone was looking at me, judging me. I was sure they were wondering how Jesus could love a sinner like me. I was literally sweating as I walked into the store. I sensed evil, confusion, and panic swirling around me, but I didn't know how to combat it. I ended up leaving the bookstore and walking home, just trying to catch my breath and recover from the incident.

....

"Good job, Confusion and Panic," chortled Satan. "Don't let Brad grow in the faith. If he learns how to pray and asks J for help, then we'll have an even bigger problem. Anytime someone calls upon the name of J to rebuke us, we are paralyzed. Don't let Brad grow, or you know what is in store for you."

"Yes, Master, we know."

....

"Gabriel, I need you to send angels to Brad. He is trying to pray, but Satan and his demons are attacking him. Brad is trying to get to know me, and I will not let his calls to me be stopped. I AM, Gabriel. I am able to thwart any attack when people pray, so get down to Brad's apartment now!"

"Yes, Lord, and I will take reinforcements with me."

....

Above my apartment that night, I believe there raged a battle of swords. Sparks flew in all directions that night as angels warred against demons.

....

"Ether," asked Confusion and Panic, "why are we facing such resistance? This Brad has only just gotten saved."

"I don't know," Ether replied, "but Prince told us not to let him pray; he knows more than we do."

....

I got on my knees beside my couch. At first, I had a hard time gathering my thoughts, but then, just as if a battle had been won, my mind opened and I could speak to Jesus like a friend. I prayed for a Bible and asked for the desire to get to know this new God who had eluded me for so long. I didn't really know how to pray, but it was almost as if the Lord himself was helping me.

Then a vision came to me. I was standing in front of my dad's casket, and I heard a voice say, "That little boy left me for more than twenty-five years, but I have never left that little boy." I knew immediately who was speaking, and I could contain my emotion no more. All the tears bottled up from all those years of stuffing my pain away finally poured out. But they were tears of joy, not sorrow; of repentance, not judgment; and of salvation, not condemnation. My slate was finally wiped clean, and I knew that I was indeed a new creation through the blood of Jesus Christ.

The Bible had always been tough for me to understand, but now that I had the Spirit of God living inside of me, the Word of God came alive. One of the first verses I underlined was Hebrews 12:5–6: "My son, do not despise the chastening of the Lord, nor be discouraged when you are rebuked by Him, for whom the Lord loves He chastens, and scourges every son whom He receives."

....

"You are finished, Panic and Confusion!" screamed Satan. "What did I say about stopping Brad from praying?"

Trembling in fear, Panic spoke for both of them. "We couldn't stop him, Prince. We tried—really we did!"

"No explanations! Out of my sight forever!" roared the evil adversary. "You have failed me far too often. Never come back to me. You will suffer for eternity!"

Then, looking around as if searching for something, he bellowed, "Depression, come here! You need to get up to speed on Brad Henry. Stop him from testifying about J to anyone. Do you hear me?"

"Yes, Master, I will do what you say," the demon fervently promised.

"A Change Of Heart"

Everything in my life and in everyone else's life is part of God's perfect plan. There are no mistakes.

Four months after I was saved, a friend contacted me to gauge my interest in starting a bottled-water company in Phoenix, Arizona. No one was holding me back, so I moved to Phoenix in March of 1993.

Phoenix was a great place for me to get back to biking and running. The weather was incredible. One night (yes, at night—I don't know what I was thinking) I set out on a forty-mile bike ride. I didn't mind riding at night because I had a light on my bike, and it worked quite well.

About two miles from home, I had to go under an overpass where the interstate crossed. As I approached the area, I saw it. "It" was a steel grate, and there was nothing I could do to avoid it. Before I could react, my front wheel caught in the grate and all went dark. The next thing I felt was a huge blow to my face and a feeling in my brain as if I had just been electrocuted. This was followed by a blast of bright light and then darkness again.

Fortunately, I quickly regained consciousness and realized that traffic was coming. I managed to roll to the side of the underpass and then pulled myself to my feet. Making my way to a nearby restaurant, I stumbled into the men's room. I had lost some front teeth, and there was gravel embedded in my chin. I went to the emergency room of a local hospital and received a number of stitches in my chin before returning home that evening. I was lucky I did not break any bones in the accident. The next week, the fun began at the dentist's office. Little did I know that this would not be the last time I would lose my front teeth in a bike accident.

Except for the bad spill on the bike, it felt great being in Phoenix. It was a fresh start in new surroundings. Now I just needed to find a good church. As God would have it, I had moved to within less than a mile of a Baptist church that would help me grow in my newfound faith. This was a church that ministered and taught the Word of God. It also had a strong Evangelism Explosion program. EE is the best course I have ever taken that helps a person learn how to share the Christian faith.

In my new church, I also needed a friend and someone who could disciple me. The Lord gave me that in Robb Williams. Robb was the singles pastor at the church and made a huge impact on my life. Not only was he a pastor to me, but he also became a great friend. This was a wonderful time for me to immerse myself in a group of godly men and learn the true meaning of the cross. Little did I know that this education was God's preparation for sending me out to minister to the world.

My business partner, a man named Jim, had known me before I became a believer, and he could now see the change in me. "Now, Brad," he questioned me one day, "I have heard some wild stories in my lifetime, but all of yours are true. What has changed the most in your life since you found religion?"

"It's not religion," I explained. "It's Jesus who has made the difference."

"Okay, whatever, just tell me why the change."

"Well, the girls I dated before, I always told them upfront that if they got pregnant, it was up to them to fix it. I had no use for any of their problems. Now I am sickened by the thought that I could have had any part in something like that, and I'm so grateful that neither I nor any girl I dated ever had to make that decision. Also, I will never have sex again unless I am married. No one told me I couldn't; I just felt the Lord telling me this in my spirit. If I am ever to have a healthy marriage, then I need to stop my immoral ways—not out of duty, but out of love for what Jesus saved me from."

Jim nearly fell over in his chair. "You mean to tell me," he asked incredulously, "that if you never get married, then you are not going to have sex?"

"Yes, Jim, that is exactly what I'm saying," I calmly answered.

"Well, I know you, Brad, and if this actually happens, then I might have to learn who this Jesus is. But my bets are you will lose!"

For me to draw a line in the sand and say I would have no more sex unless I was married could come only from God. But God is in the life-changing business, and he was surely molding my heart into the heart of one of his followers.

Sexual immorality had been my lifestyle for over twenty years, but I began to search the Bible for passages on the topic. This sin would be a hard one to break, I knew. I diligently looked for answers in the Bible and prayed consistently about the issue. I remembered a few years earlier when I had sat on an airplane beside a woman in her seventies and without the slightest bit of remorse or conviction pulled out a Playboy magazine and looked at the pictures. Never once had it dawned on me that this behavior was not right. In fact, I was just waiting for the elderly woman to ask me to put the magazine away so I could confront a "self-righteous" person. But now, recalling the incident, I was appalled at my behavior.

In the past, whenever someone would try to trick me in my business dealings, I would spew a volley of swear words that would make a sailor blush. Once a person kept promising to send me a part that I needed for a big shipment, but he never came through with his word, causing me to miss my deadline. When I discovered that he had been lying to me all along, I swore and ranted and raved, threatening to tear his heart out with my bare hands if ever I laid hands on him.

Now, in small steps and ways, my business methods and my approach to people were slowly changing. Marveling at the difference, I kept asking myself how this was happening. However, now that I was on the side of the King, I mistakenly assumed that my business would automatically boom, but in reality, things only got worse. I would go on a road trip and line up a lot of bottlers, for example, only to see most of the orders vanish into thin air once I got home. In my prayers, I often asked God, “Am I still in sin?” or “What am I doing wrong?”

....

“Brad asks a good question, Jesus,” said Gabriel. “If you are on his side, why is his business not successful?”

“First of all, Brad is not ready for wealth yet,” answered the Savior. “He is just a baby in the Word and in this new way of life. I don’t want him to be distracted by earthly things, as his heart is not ready for that. Over time, he will come to know that I love him more than he could ever imagine and that I will provide for his needs, not his wants.

“As you know, Gabriel, John the Baptist testified of my coming and preached repentance. He did everything right and nothing wrong, yet he was ultimately beheaded. Earthy pleasure and riches, as well as a safe Christian life, are not guaranteed, Gabriel. In fact, it is usually just the opposite. Many people suffer for following me. They are targets of the evil one and suffer persecution from him, but I will use their circumstances to make sure that they grow in the faith. One cannot grow in faith in heaven, Gabriel; that happens only on earth.”

“So why does man do good works while he is on earth, Jesus?”

“Man is saved by faith in me alone, Gabriel, and not by any good deeds. If man could enter heaven based on his good deeds, then I suffered, died on the cross, and conquered death for nothing. But it is exactly because man is powerless in his sin that I needed to become the perfect sacrifice. Man should do his good works out of love and thankfulness for what I have done for him— nothing more. Faith, Gabriel, comes from the heart, not the head. Sometimes that distance of twenty inches from the head to the heart keeps people from enjoying eternal life with me.

"The reason most people on earth are not happy is because they want earth to be heaven. They want life on earth to be easy. But no one on earth will be happy, Gabriel, unless they can yield everything in their life to me. Treasures on earth do strange things to the human heart, don't they Gabriel?"

"Yes, Jesus, you would think that after a while, people would get it."

"I am the only one who can understand the human heart, Gabriel. Satan knows he is defeated and has nothing to lose, but he has hooked many men and women into believing his lies. One of his biggest lies is that happiness is found in things. There are many ruined lives over this idol, Gabriel.

"Will man ever get the true meaning of life? Not on this side of heaven, Gabriel. It is only by faith in me and the things unseen that anyone will inherit eternal life. The minute the elect see me in heaven, they will finally understand."

....

Hope

"Pastor Robb, tell me about Julie, that girl who works for you," I asked my friend one day. Julie had earned a master's degree from Denver Seminary and knew much more than I did about the Bible. Plus, she was cute.

"Why? Are you interested?" Robb asked.

"Well, maybe. Do you think she would go out with me?"

"Only one way to find out, Brad," Robb said, smiling as he spoke.

A short time later I called her at the office: "Julie, hi, this is Brad Henry. How are you?"

"Great," she replied. "Are you looking for Robb?"

"No, I was . . . umm . . . ahh . . . well, do you like Mexican food? Would you like to go to Carlos O'Brien's with me?"

I would later find out that I had caught her off guard and that she said yes without really thinking. Nonetheless, we enjoyed a great dinner together. It was the first time I had ever been on a date with a fellow follower of Jesus. Towards the end of dinner, I asked Julie if she would like to go out again. She replied, "Yes, that sounds good. Maybe we could go out with a group of people." Now I had been in sales long enough to recognize a brush-off. It was not the answer I was looking for, but at least we had enjoyed a good dinner.

Every time I called Robb at his office, Julie seemed to answer, and we were always cordial to each other. The Fourth of July was drawing near, and Robb and his wife, Sarah, were coming over to my house for the day.

"Hi, Julie. This is Brad. Is Robb there?" I asked when calling one day.

"Yes, I'll get him for you," she responded and then added, "What are you doing for the Fourth, Brad?"

"Oh, I'm having Robb and Sarah come over to the house," I casually remarked.

"Wow, that sounds like fun," replied Julie.

Now I had been in sales long enough to recognize a hint when I heard one. Seizing the opportunity, I blurted out, "Hey, Julie, would you like to come over and join us?"

"Sure. That would be great!" she quickly responded. "What time?"

I gave her all the details and then hung up the phone, more than a little perplexed and quite excited. Was she really coming to my house?

After the Fourth of July, Julie and I became inseparable. After a couple of months of dating, I knew Julie was the one I wanted to spend the rest of my life with. On my birthday, we drove to a restaurant on the top of a mountain that overlooked Phoenix. After dinner I proposed, and—whew!—she accepted. I did not have an engagement ring that evening, but I had an idea of what Julie would like.

A few weeks later, I was ready to surprise her with a ring. Since my twenties, I had suffered from migraines that were at times debilitating. Julie knew I always kept a bottle containing Advil and other medicines nearby. I put the ring in the Advil bottle, and on our way out to dinner, I said to Julie, "Wow, it feels like I have a headache coming on." Julie immediately got the bottle out, removed the cap, and pulled the cotton out so fast that the ring flew out and landed in her lap. She didn't even realize it at the moment, but after she handed the Advil to me, a strange look crossed her face as she slowly looked down and saw the ring resting on her dress. And oh, did I mention that the window was down while all this was taking place? Fortunately, the ring did not fly out the window, and my friend Robb married Julie and me in November of that year.

The Lord knew it was time for this great blessing in my life. He had delivered me unto himself and protected me from myself. He had set me free from all my addictions and restored my life. Now he gave me an earthly treasure that would stand firm with me in the faith, and together we would walk this new journey down the road of life.

....

“Depression, we have a problem. Brad now has a praying wife. Send Frustration, Dismay, and Darkness to thwart his business efforts. Keep up the good work of dangling hope and then at the last minute taking it away. Even though Brad has put his faith in J, his identity is still in what he does for a living. I want him destroyed in that area so that he is an ineffective witness, even to his new wife.”

“Consider it done, Prince” came the usual answer.

Will Business Ever Change?

After two years of the major bottlers watching our sales and our competitors, the rumblings were that Pepsi and Coke had realized how good the market was and were going to introduce their own brands of bottled water. The writing was on the wall for us.

"This bottled-water business is getting tougher and tougher," I explained to my partners. "The bottlers that distribute for us want to start producing their own water. Now they see it as a viable source of income." As a group, we decided to sell to a larger company.

I found an interested company that liked the way we had built the business. Meshing the two companies made sense. I met with the president of the new company, and we started all the paperwork for due diligence. The problem was that the president did not have the final say; that lay with the owner, a wealthy and very shrewd businessman.

"Great news, Julie," I informed my wife. "They said we will get the deal done by next week."

"Brad, we are so short on money. I pray that this goes through," she worriedly responded.

"I know; it has to. We've had so many ups and downs that it's been like a roller coaster," I replied. Then, praying silently, I said, *Lord help us to sell this company and take care of the debt we owe our investors. Help us to have a fresh, clean start. I pray that we would have favor with everyone we are dealing with and that you would prepare the hearts of everyone in the meeting.*

The day to sign the paperwork came, and we met on the thirty-fifth floor of the new company's attorney's offices. The proposed deal would pay off all my old investors and pay me back all the money that was owed to me. Julie and I had not taken a paycheck in over nine months, so we were due approximately fifty thousand dollars. This would give us a fresh start, one that we desperately needed.

The owner looked at me and said the following: "Brad, we like what we see in your company. The president has told me that this will be a good fit and that you will help him with all the beverage distributors that you have used. So this is the deal: I will pay your investors all the money they are owed, but I will not pay you any of your back pay. Take it or leave it."

Julie and I had enough money for only one week. But as I looked across the table, I knew I desperately wanted to be clear of this debt to the investors. So I signed the papers and then went home to break the news to Julie.

"Julie, they took the deal under one condition: we will not get any back pay."

"How can they do that to us? You worked so hard at this!" my wife protested.

"I know this is tough to take, but at least we are free and clear of the debt. Plus, I still have a job with the new company."

"But we have personal debt from not taking a paycheck," Julie reminded me. "How will we pay that off?"

"I'm not sure, and I'm tired, just like you are. Sometimes I wonder if we'll ever get a break," I sighed.

"I don't know either, Brad," Julie replied. "This is the first deal I have been through with you, but you've told me all the horror stories of deals in the past that just evaporated. Are we ever meant to get ahead?"

My first day of work, I met with the president of the company, and he told me that they would like to go after the Major League Baseball license. The plan was to secure the rights to all teams and use their labels on the bottles. In New York, the distributors would feature the New York Yankees or New York Mets, depending on the region of the city. In Baltimore, a distributor would feature the Baltimore Orioles, and on and on. For the next eight weeks, I flew across the country, setting up new distributors and informing my existing distributors of the new product line. They were all excited and started to order truckloads of Major League Baseball licensed water. We soon had distributors reordering, and things looked very promising.

"Brad, you have a call on line one," my secretary informed me one day. "It's Major League Baseball."

I quickly picked up the phone, and the voice on the other end of the line greeted me. "Brad, how are you today?"

"I'm great," I responded, "and thank you for granting us the license. Sales have been really good."

"Well, that's why I'm calling Brad," explained the voice. "It seems that when we granted you the license, our sponsorship and licensing department got their agreements crossed. We inadvertently granted you a license although one of the major soft-drink companies is already paying us millions of dollars a year to be the official beverage of Major League Baseball."

"But we aren't selling soda; we're selling water," I protested.

"I know, Brad, but in the agreement with us, they put in the clause 'all beverages.' You have four weeks to sell off what you have, and then your license will be terminated."

"Wait, do you know how much money we have put into this?" I asked, incredulous at the direction this conversation had taken. "We have approximately $250,000 in capital and marketing invested in this. How are we going to recoup that, and besides, what will our distributors think?"

"I am sorry, Brad," the man said, "but the best we can do is to give you twenty-eight tickets to the World Series next month. Four tickets to each game."

"Let me talk with our president, and then I'll get back to you," I managed to get out.

"Okay, Brad. Just let me know."

Well, how many things in life look like they are going to turn out great but then end up kicking you in the teeth? That's exactly where I was at the moment.

"Ron, we have a problem," I began when talking with our president.

"So what's new, Brad? We have problems here every day," Ron laughed.

"Major League Baseball gave us rights they didn't have, and now they need to pull the license from us."

"Get the attorneys; we'll sue them" was the quick reply.

"Now, Ron, I agree that we could do that, but if we ever want a license again for anything in baseball, we would ruin our chances by suing them now."

"Okay," Ron agreed. "It's in your hands, Brad. Just make sure we don't lose our distributors."

My calls over the next couple of days to the distributors were horrifying. I got called every name in the book. When you sell to a distributor, they buy in truckloads. Then they have a couple of thousand clients that they sell to in amounts ranging anywhere from one to hundred cases. Now these distributors would have to go back and tell their clients that they could no longer offer our product. The trickle-down effect was enormous.

My only saving grace was that our biggest distributor was located in New York, where the World Series was being played. So I offered them tickets to the World Series. Things were looking up somewhat, as they could tell their kids they were going to the World Series. Some distributors even said that this was the best present their kids had ever had. Things were good, it seemed.

Game one of the series was scheduled for October 19, 1996, at Yankee Stadium. However, Major League Baseball wouldn't know the final teams for the game until ten days before the event. Only then could the tickets be printed, which meant I would not get the tickets into my hands until three days before the game. In the meantime, distributors were angrily demanding the tickets, warning me this better not be a sham to keep their business. Yikes! Could things get any worse?

As I said, game one was scheduled for October 19, a Saturday. On Thursday I got the tickets, which meant I had to FedEx them that night so that the people could have their tickets by Friday for the game on Saturday. I made the arrangements and called my distributor to give them the tracking number, and out the tickets went for Friday delivery. Whew!

All of a sudden, I got a call about a nor'easter moving into the region. The weather report from the NOAA (National Oceanic and Atmospheric Association) warned of heavy rainfall up to twelve inches in some places and wind gusts as high as eighty miles per hour. *No, no, no!* I immediately called FedEx and listened with a sinking heart as they informed me that their flights would not be able to deliver the packages on time. In fact, the weather was so bad that game one was called off and rescheduled for October 20.

The problem was, my distributor still did not have his tickets. Hesitantly, I picked up the phone and made the call. "Bill, I have some bad news. Because of the weather, FedEx is not able to deliver your tickets."

"My son has been counting on this! Do you know what you have just done to him?" came the understandably angry reply. "What are you trying to do? We're finished if you cannot get me my tickets!"

Well, I will admit that I had had so many great prospects and deals that just vanished that I was used to this crazy outcome. Still, in despair I silently prayed, *Lord, why? Am I doing something wrong? I don't know how much more of this I can take.*

I called FedEX and in my most polite, calm, and persuasive voice threatened to sue them if they did not find my tickets. Not only couldn't they deliver the tickets they could not locate where they were. Usually Fed Ex knows where a package is but they had no clue. Talk about frustrating. "Do you know what you just did to our company?" I tried to explain.

"I am sorry, sir, but the maximum liability for each package is one hundred dollars."

“One hundred dollars? Are you kidding me? We just lost $250,000, and all I can get is $500 for the packages lost?”

“Yes, I am sorry, sir, but that is our policy.”

Wow, it was tough being a new Christian. Some good swear words would have been appropriate right then. No, maybe not appropriate, but they sure would have conveyed what I was feeling. Thankfully, the Lord somehow calmed my nerves, reminding me that though life is full of surprises, God is still in control. I was being tested, and it certainly wasn’t fun. But at least I did not swear.

Even though I did not swear, when I got back on the phone and told my distributor that Fed X could not find the tickets to either game one or game two because of the weather he used up all the swear words I could have used. But now that I was the father of a baby boy, I could understand his pain. To tell his son he was going to the World Series, and then for that boy to tell all his classmates that he was going, and then for the whole thing to fall apart was horrible. I had to take the fall. It was my fault. Blaming others would not fix the situation, but they needed to blame someone, so it might as well be me.

But God does the unexpected all the time for our good. It turned out that if our distributors had gone to either game one or two, they would have been there when the Yankees lost. However, because of the mix-up, Major League Baseball gave us tickets to game six, if there was to be a game six. Now I had to pray that the Yankees would not get swept in Atlanta. Fortunately, the Yankees won the next three games in Atlanta, which meant they were coming home to New York to play game six. That meant our distributors were happy again, which meant I was happy again. Then God gave us favor and the Yankees won game six at home, with our distributors and their kids there to see it happen. In the end, we kept our distributors, and they were one happy bunch.

In life, issues will always crop up that are beyond our control. The bottled-water company was having some problems with product shipped to Japan. This composed 70 percent of our business, so we knew that in the end people were going to have to be laid off. Six weeks later, Christmas was right around the corner, and it looked like the company was not going to make payroll. Realizing I needed to find another job, I gave my two-weeks notice. The company kept sputtering for another year before it finally closed its doors.

I interviewed for a position to sell water-purification systems to homes. The system was expensive but worked very well. Julie and I had no money in the bank, so Julie’s parents helped us get through the next month. I have to say, I have awesome in-laws. They thought their daughter’s husband had a stable job, but little did they know—or Julie and I, for that matter—what was about to come.

My first night out to visit prospective clients was a rude awakening. As I visited in some of the homes, I realized they did not have the money for such an expensive system. Part of the sales pitch was for me to use a ploy to get people to put the purchase on credit, even when they couldn’t afford it. But as I looked into the eyes of the prospective clients, I could not in my heart urge them to go into debt for this purchase. Instead, I told them the truth, that they should not get the system, and quit that night.

When I arrived home that evening, I told Julie what had happened, and we both sat on the sofa and cried. I called a great friend of mine, and he came over to pray with me. As he prayed, we felt the Lord’s presence and peace, but the floodgate of tears opened again. This constant up and down and back and forth in our lives for so many years was taking its toll, though we did not realize it.

What was I going to do? I was an airline pilot whose career had been cut short. I had started numerous businesses that had all looked promising but then ended under odd circumstances. I had all this training and experience, but for once in my life, I started to doubt my ability. But maybe that is where God wanted me. Maybe he wanted me to trust him and not myself anymore. I could agree with that revelation, as nothing I had done so far in my life had worked out. But that still posed the question, what now?

A Life-Changer

After a year of marriage, Julie and I wanted to start a family. Within a few months, Julie was expecting.

Julie was six months along when I got a call from her while I was traveling. The doctor had performed an ultrasound and discovered that our baby had cysts on its brain. With her words, I immediately thought, *This can't be happening!* But Julie went on to say that the doctor also said that he had seen this many times before and that the cysts always went away before the baby was born. *Then why did they even bother to tell us? This is crazy. What was supposed to be another three months of joy will now be three months of wondering what will happen,* I silently fumed.

Julie and I knew we had a choice. We could either worry ourselves sick or pray and trust the Lord for the outcome. And pray we did. Then, on August 26, 1996, nine days short of his due date, our son Bryce was born, healthy and normal in every way.

I was forty-one years old and had no clue of how to be a father. I had never even changed a diaper, but better late than never. What amazed me the most was the fact that the hospital let Julie and me leave with this little baby. Shouldn't a nurse stay with us for a couple of years? But the Lord does things by small degrees. Little by little, Julie and I learned how to take care of Bryce. He was our treasure. Through him I began to see love in a whole new way. Now I had a glimpse of how much God the Father loves his children.

Two and a half years later, Julie and I were blessed with another baby boy. Chase was born on March 15, 1999. By then we were feeling pretty good about our parenting skills, but no one could have prepared us for what was to come. For about the first six weeks after Chase's birth, Julie and I felt as if we had ten kids instead of two. But we slowly got the hang of things and finally adjusted. Julie was even able to synchronize the boys' afternoon naps every now and then. And as anyone with two small children knows, that is nothing short of a miracle.

Now we had our family, but I still needed a secure job.

NASCAR 101

Have you ever felt that God was silent? Have you ever felt as though you were in a desert with no oasis in sight for hundreds of miles? That's how Julie and I felt during that time when I needed a job. We often prayed, "Lord, what do you want us to do?" and waited for an answer.

As I looked back over my life, I felt that everything I had ever loved had been taken away. Of course, some of that was the result of my own stupidity, especially that one joint I had smoked, which began the panic attacks. If I had not smoked that joint, I would have still been flying. But then, if I had still been flying, I may have never come to know Jesus. I would not have met Julie and on and on. We can't outrun destiny.

When I had owned various companies, I depended a lot on independent sales representatives in other parts of the country to introduce our products to prospective buyers. They were not on our payroll, but we would pay them a percentage of the sale if they sold our products. After my last business venture with the bottled-water company, the last thing I needed was inventory and a payroll, but why couldn't I be a sales rep? It just so happened that Circle K was headquartered in Phoenix, along with a number of grocery chains. All I needed were some manufacturers to represent, and then I could sell their products to the stores in the Phoenix area, I thought.

I started making phone calls, and in four weeks, I had five companies that I was representing. One of the companies was run by an investor in the back-cushion company I had owned. I made a sale for them, and my commission was approximately $3,500. I called to see if we could get our check, and they said it would be no problem. Julie and I were ecstatic. Finally—a break to make a living again, with no overhead!

After a week, however, the check still had not arrived. I called the vice president of sales, and he explained, "Brad, the owner of the company says that you owe him five thousand dollars from his investment in your old company. He is going to keep your commission and apply that to your account." Now if he had said this upfront, that we still owed him money and needed to work it out this way, that would have been fine. But instead, it felt much like, *Here we go again! Does anything ever turn out for us as planned?*

When I was growing up in western Pennsylvania, we often went deer hunting at night. We would spot the deer by shining a light into the surroundings. When the deer saw the light, they would freeze in their tracks and just stare straight into the light, their eyes glowing red in the night. From that practice comes the saying "looking like a deer in the headlights." Well, when I told Julie, who had been waiting for the check just as eagerly as I, about the problem, it was as though I had shined a spotlight on her. She just sat motionless and speechless there on the couch, frozen in the terror of the moment.

The only thing I knew to do was to let it go and keep working on securing commissions from other companies. I did manage to sell some product, but it was barely enough to get by. Meanwhile, Circle K signed an agreement to become the official pit stop for NASCAR products. The buyer at Circle K said that I would have a better chance of selling to him if my products were licensed with NASCAR. An investor in our cushion company had taught me the licensing business, and I had certainly learned a thing or two from the bottled-water fiasco with Major League Baseball.

The licensing business is all about negotiating to obtain the rights to use a certain character, like Spiderman, or specific property, like NASCAR, on a product. For negotiating those rights, I would get paid a small percentage from everything sold under that license. So after talking to the Circle K buyer, I immediately made a call to NASCAR in Charlotte, North Carolina. I was expecting to get some automated corporate machine, but much to my surprise, a great guy answered the phone. He was so helpful that I wondered if he was a believer. I didn't get the product licensed, but this man became one of my best friends.

We then signed on another company to represent. This time we got licensed with NASCAR and all the major race teams. We were in business.

Then my friend in NASCAR said that he had a friend in Phoenix that he would like me to meet. At the same time, a partner was providing me with capital to stay afloat. This man not only got me started in my business but has also given me more great advice over the years than anyone else I know.

Anyway, I met Reed, the man in Phoenix, and we got along so well that for the next thirteen years, we were like brothers. We turned our Sales Rep company into a Licensing Company. Reed had a friend at Universal Studios, and a huge licensing deal with them really started our company on the right track. In fact, we eventually had so many clients that either Reed or I needed to move to Charlotte, North Carolina, where the NASCAR headquarters were located. I spoke with Julie, and we agreed that it would be a good move for us. I am from the East Coast, so it was great to move closer to home.

Over the next ten years, with a couple of great partners, we started licensing products into the NASCAR community. At one point, the Lord allowed our company to be one of the largest licensing companies in the sport of NASCAR. Our clients were doing over seventy million dollars a year in licensed business.

Now that business was going so well, I wanted to focus on getting back into good shape. Maybe I could try to do a full Ironman.

Panama City Failure to Full Ironman

I wanted to start training again for a triathlon, but my teeth had been bothering me ever since my first bike accident. I went to a dentist, but he did not have good news. I needed five root canals and six crowns. Now, I would rather have bamboo shoots forced up my fingernails than go to a dentist, but here I was, obviously facing numerous dental procedures.

On my third visit to the dentist, the plan was to do two root canals. After injecting me with Novocain, the dentist started to remove the root. The pain was as intense as if I had received no Novocain at all. The dentist gave me more of the anesthetic, but still there was no improvement. I received five more shots, and as I waited to see if they would take effect, I heard a guy in another room let out a howl. I thought, *Good, at least someone else is also suffering.*

The dentist finally returned, and he was apologizing. "Brad, I'm so sorry, but apparently I gave you an outdated dose of the medication. I can give you 10 percent off the bill."

Are you kidding me, Doc? I just got all these shots of Novocain and all I get is 10 percent off? Okay, I didn't say that, since I knew he still had more work to do, and I didn't want him to get mad. But I was certainly thinking it!

Finally, four weeks after the dental work started, I was finished. I was never so glad to leave a dentist's office in my life as I was on my last visit. I was even thinking, *Maybe I should never ride my bike again. I don't want to ever have to go through this again.* But I have an impulsive personality—shoot, then aim. Just as quickly as the first thought had come, a new one took its place: *Better start training. I'll worry about my teeth later.*

I couldn't shake a nagging feeling about the Panama City Half Ironman I had competed in ten years earlier. I had not finished and certainly didn't want to end on that note. Now that my business was doing well, I wanted to do something about my body, which had gotten out of shape.

I am a person who needs to have a goal in order to make myself do something, so I asked Julie, "What do you think about me doing the Ironman at Lake Placid, New York, next year?"

"Next year?" Julie asked. "You can only run a few miles now!"

"I know, Julie, but I need a goal."

"Brad, can't you ever do anything small?" my wife laughed.

Well, I guess not. It was now October of 2004, but I signed up for the Lake Placid Ironman for July of 2005.

I set up an Excel spreadsheet and started to log my miles. In the beginning, I could swim five hundred yards, bike twenty-five miles, and run three miles. But I would need to swim 2.4 miles, bike 122 miles, and then run 26.2 miles. The event seemed so far off, however, that I didn't worry—at least not yet.

By April of the following year, I had signed up for another Half Ironman in White Lake, North Carolina. My mileage was sufficient to complete the race, but in any long-distance event, you never know for sure how your body will react. At least this time the swim would be in a lake, not the ocean.

The morning of the race arrived, and the weather was great. I completed the swim on schedule, and I finished the fifty-six bike miles in about three hours. The run of 13.1 miles lay ahead. As I began the last leg of the event, I could not get Panama City out of my mind. My legs felt like lead, but I started praying and asked the Lord to help me run this race. At mile three, my legs started to loosen up, but by now the temperature had started to rise. At mile eight, I began to wonder if I had another five miles in me.

Doubt is a horrible disease. It can paralyze you and keep you from being the best that God has created you to be. Doubt can create anxiety and rob you of joy. I was experiencing all that, but I prayed again and the doubt subsided. Then, at mile ten, my strength was suddenly renewed, and I ran the last three miles faster than I had run the first three. There it was—the finish line! I ran the 13.1 miles in one hour and fifty-two minutes. What a feeling to accomplish something that I had quit at years earlier!

Instead of relishing in my finishing, however, I now began to wonder about the upcoming Lake Placid event. If I had struggled with a Half Ironman, how would I ever do an entire Ironman that was just three months away? My solution was to step up my training. On weekends I would do a ninety-mile bike ride followed by a three-mile run. Then the next day I would do a long run from twelve to fifteen miles in order to simulate running on tired legs.

Time passed. I needed just one more long bike ride before Lake Placid, and then I could start reducing my mileage to get ready for the race. During the next six weeks, I built up my mileage and then dropped back down to normal. Two weeks before the race, I started to get nervous. The dream that had started in Panama City was about to become a reality.

Lake Placid was a beautiful place, but I began to wonder about the wisdom of my tackling the Adirondack Mountains. Why hadn't I picked a flat course? Nonetheless, the morning of the race was quite exciting. There were over two thousand of us with one singular goal in mind—to finish. If I could complete the race quickly enough, I wanted to try to qualify for the Ironman World Championship that would be held later in Kona, Hawaii.

The cannon sounded at 7:00 a.m., and the swimming portion of the event began. We looked like a bunch of carp fighting for bread, with everyone swimming all over each other. It was mass chaos. After about five hundred yards, however, the crowd thinned out, and the long day stretched ahead.

I have written many stories about the Ironman and life, as there are many similarities between the two. The Ironman is a long race in a long day. If you are in the swim portion, you can't be thinking about the bike ride and definitely not about the marathon. You have to think about what is in front of you at the moment, or you will defeat yourself in your mind. The same holds true with life. How many times does a trial come up and we think of all the bad things that might happen instead of focusing on what is in front of us at the moment? Most of us lose our joy in life by looking too far down the road and missing the miracles right in front of us.

The swim portion of the Ironman was in beautiful Mirror Lake. The lake was not large enough for us to swim a mile out and back, so the swim course was composed of two 1.2-mile loops. At the end of 1.2 miles, we had to exit the water and then enter again at the same place where we had started for another 1.2- mile loop. I tried to remain in the moment. *Think only about the next stroke . . . the next breath . . .* I reminded myself, and on and on. Finally I exited the water on the second loop at a time of one hour and fifteen minutes, an average time for most of the contestants. After removing my wetsuit, I ran the four hundred yards to the transition area where my bike was stored.

Someone was waiting for me with my bike. I jumped on, and off into the mountains I went. The course was very mountainous and consisted of two fifty-six-mile loops. The last fifteen miles of each loop ascended Whiteface Mountain. Why couldn't it have been the first fifteen miles? At the end of the first loop, my time was three hours and fifteen minutes, which was pretty slow. However, as I entered the town of Lake Placid, all the people were cheering and my spirits lifted.

That rally was short lived, and I was soon headed out of town for the last loop. At mile eighty-five, my mind began to wander. Big mistake! I was getting tired, or was I allowing myself to get tired? I had now been exercising for approximately six and a half hours. I was getting a little light-headed, knowing that I still had to climb Whiteface Mountain again and then run a marathon. *Are you kidding me? Run a marathon?* I remember that I saw a couple sitting in a mobile home and watching TV, and I desperately wished I could go inside and sit down with them and call it a day.

But each time I thought I could go no farther, I prayed and found new strength. I also knew that quite a few people were praying for me to complete this race, and that gave me strength to continue. It wasn't a matter of qualifying for Kona anymore; it was simply a matter of survival or whether I could even finish the race.

As I rode, I kept seeing a sign with a quote from Lance Armstrong: "Pain lasts but for a moment; quitting lasts forever." I wanted to quit so many times, but I kept seeing this sign. I was soon ready to beat up the person who had placed it in so many strategic places. Then I thought to myself, *Well, this pain isn't for a moment; this pain is lasting a long time!* In life we can justify anything in order to avoid the pain involved from hanging in there. But after praying, I knew I needed to keep going.

I completed the second loop of fifty-six miles in three hours and thirty minutes. My combined total of six hours and forty-five minutes was a very slow bike time. When I dismounted my bike, my legs felt dead and heavy, like telephone poles.

I walked into the changing tent and put on my running clothes. However, the longer I sat in the tent, the longer I wanted to stay there. I knew that a marathon lay ahead of me, and I was not eager to start it. But I also knew that the longer I stayed in that tent, the longer it would take me to finish. Instead of my transition taking the usual five minutes, it took about twenty.

I finally mustered the nerve to start the run. As I passed under a sign that read "Run Start," I knew it was now or never. After I had run about three miles, my legs started to come back to life. The run course consisted of two loops of 13.1 miles each. The first loop went pretty well, and I finished it in two hours. As I was heading out of town to run the second loop, I hit a wall. By the time I reached mile fifteen, I was walking through the aid stations.

At the mile-seventeen aid station, I didn't think I could keep going, because I was starting to cramp severely. During the Ironman, participants don't drink any water because that would dilute their salt levels too much. In the bike portion of the competition, we ate hard foods, such as Power bars or bananas, but during the run portion, we took in only liquids. The aid stations were supplied with energy gels, Gatorade, chicken broth, oranges, and, best of all, decarbonated Coke, which has a ton of sugar and caffeine and supplies quick energy. But what I needed now was salt—and quick. I could not seem to replace the salt I was losing quickly enough.

The entire race was 140.6 miles, and I had now completed 130 miles. Because of the cramping, I doubted whether I could go on. As I walked into the aid station, I prayed silently to myself, *Please, Lord. I need salt.* As I looked around, I saw an elderly man with a Morton salt container in his hand. He instructed me to hold out my hand, and then he poured pure salt into it. I guzzled it down, along with a Gatorade, and thought, *God sure answers prayers quick!* Within a minute, I was able to start running again, and I soon made it to the mile-twenty aid station.

At the aid station, I looked around for my friend with the salt. "I need salt. Where's that man with the Morton salt?" I asked.

Perplexed, the people at the station stared at me and said, "There's no man here with salt."

"But he was at the other station just forty-five minutes ago. Why isn't he here?" I protested.

"Sorry, but there's no one here giving out salt."

I know that on that day, more than a thousand people were praying for me. Our fight is in the heavenly realms as much as it is in the physical realm, and sometimes God assists us in supernatural ways. As Hebrews 1:14 says, "Are not all angels ministering spirits sent to serve those who will inherit salvation?" If the Lord could send an angel with Morton salt to assist me on an Ironman course, he can send an angel to help you today in whatever circumstance you face.

I tried to drink as much of the decarbonated Coke as I could, but I was still cramping. As I approached mile twenty-three, I could see in the distance the hockey rink where the "miracle on ice" occurred when the Americans beat the Russians in the 1980 Winter Olympics. I could even see the home stretch, but for some strange reason, I still was not sure I could make it.

All of a sudden, I passed a woman with gravel on her back and the side of her face. Apparently she had gotten dizzy and fallen down an embankment. She was bloodied, and I asked her if she needed help. She just shook her head no and smiled, and I marveled at her resilient spirit that refused to give up.

I had three miles to go and was now growing concerned that the cramping would prevent me from finishing. Every two hundred yards, I had to stop and try to stretch out my calves and hamstrings. But I was in town now, and people were lining the streets cheering us on. All I could think about was Julie, Bryce, Chase, and that elusive finish line.

At mile twenty-five, there was a short but incredibly steep hill to climb. All of a sudden, I heard the fight song from the movie *Rocky* and was able to make it to the top of the hill. I had one mile to go. I am sorry to say that even at this point, my mind would not allow me to feel joy that I was about to finish. I still doubted whether that would happen, even though I was so close.

Time blurred, and then, four hundred yards in front of me, I saw the Olympic oval that marked the finish. The crowds were six deep, and I started to cry. I heard one person yell, "Hey, look. He knows he's going to make it!" When I had only two hundred yards left to run, Julie, Bryce, and Chase came out to run the last yards with me. We held hands, and all the pain evaporated. We were coming home.

At 7:56 that evening, twelve hours and fifty-six minutes after the start of the gun that morning, I broke the tape and finally stopped. I could not hold back the tears, but I also could not keep my balance. Fortunately, Ironman assigns people to help each participant at the finish because they know what the body might do. A volunteer helped me walk for a few minutes to make sure I was okay. Fifteen minutes later, I was back with my family, celebrating my finishing. We all rejoiced together, because my wife and sons had sacrificed as much as I had in order to accomplish this goal. They had spent numerous weekends allowing me to go out on long bike rides, picking me up at the hospital, and sacrificing their time so I could do this. It was a family effort and a family victory.

When I crossed the finish line that day, I remember thinking that heaven would be like this. Most of us are tired, and some of us think that we can't go on. We wonder how we can make it another hour, much less another twenty, thirty, or forty years. But just like in the Ironman, we can't allow ourselves to dwell on how far we have yet to go; rather, we must focus on what is in front of us to do right now. God will take care of the rest.

At the finish line of the Ironman, a famous announcer, Mike Reilly, waits to say your name, followed by "You are an Ironman." If we continue to keep up the good fight, if we keep on keeping on in the faith, if we refuse to give up, then one day at the finish line in heaven, we will hear the voice, not of Mike Reilly, but of Jesus Christ himself, saying, "Well done, good and faithful servant" (Matthew 25:21).

Ironman Wisconsin

I started feeling so good about my accomplishment that I wanted to do another Ironman. I thought that if I got a coach and really trained hard, then I might be able to qualify for the world championships in Kona, Hawaii, in October. In order to qualify, however, I would have to be in the top two or three of my age group.

If I could do the swim in an hour and fifteen minutes, which was my time at Lake Placid, the bike ride in six hours, and the marathon in three hours and forty-five minutes, I might have a chance. But the Ironman is a very unique event. Even the best training does not guarantee success. Many other factors, such as injuries, nutrition, race-day weather, and mental outlook play a role. Most defeat in the Ironman comes from the mind, but I had learned to do what I was capable of doing and then let the other stuff work itself out.

I now stepped up my training and implemented much the same routine as the one I had done in preparation for the Lake Placid Ironman. On weekends I did a ninety-mile bike ride, followed by a three-mile run. Then the next day I ran twelve to fifteen miles in order to simulate running on tired legs. Again I needed just one more long bike ride before reducing my mileage to get ready for the race.

"Okay, Julie, I should be gone about six hours. I'll take my phone with me. At least this is the last long bike ride before the race. Thanks for allowing me to do this. You have been awesome, allowing me to train and keeping such a great attitude. I could not have done this without your support."

"Just be careful out there, Brad," Julie answered.

"Hey, I'm always careful," I lightly replied.

I left the house at 7:30 a.m. The temperature that day was forecast to rise to ninety degrees, but at least I was starting early. Around mile twenty, I was drinking a bottle of Gatorade, well aware that I needed to consume approximately 350 calories an hour in order to be able to complete the race. As I held the Gatorade bottle in my right hand, I approached an intersection, which by now was out in the middle of nowhere. The driver in a car at the intersection looked up the road but not back toward me before pulling out.

On a long bike ride, you tend to daydream and think about all sorts of things. I always like to visualize the way an event or even life should go. I was daydreaming that day about my previous race at Lake Placid.

"911. What is your emergency?"

"I just saw a bicyclist flip over his handlebars and hit the pavement with his face. I am not sure if he is alive." Two older ladies were driving behind me that saw the accident and made the call. "He's just lying in the middle of the road, and he's bleeding pretty badly.

"Mam, an ambulance is on the way. It should be there shortly."

Then, back at my house, the phone rang. "Is this Julie Henry?" asked an unfamiliar voice.

"Yes, it is," my wife answered, wondering who was calling.

"I just wanted to let you know that Brad should be okay."

The ambulance is on its way, and they will be taking him to the hospital," explained the unknown caller.

Confused and frightened, Julie responded, "No. Brad just went on a bike ride and should be back in a couple of hours."

"I'm sorry, but he has had a bad accident," a man explained.

"No, this can't be happening. I never thought he would get hurt!" Julie replied, panic and fear threatening to take over.

"They will be taking him to Lake Norman Hospital. You can meet him there."

"I was beginning to wake up and a friend of mine who happened to be a Physicians Assistant luckily was also riding in a large group behind me. They had come upon my acident and waitied till help arrived."

I was sitting on the road and felt someone pouring water over my head. I could hear sirens in the background, but it still did not register why I was sitting in the middle of the road. "Just be still," said a woman trying to help. "You have been in a bad accident."

"What am I doing out here? How did I get here?" I asked, bewildered. I looked around and saw approximately ten cyclists who had been in another group. They had stopped to block traffic so I would not get hit by another car. One of the cyclists gave me a phone to talk into, and Julie was on the line.

"Brad, are you okay?" asked my worried wife.

"Hi there, Julie. Umm . . . ahh . . . I don't know what's going on. How did I get out here?"

"Brad, you're scaring me. Are you all right?"

"I better go, Julie," I mumbled, still not sure of what was going on.

"Wait, wait! . . . Hi, Julie. This is Scott from the development. Brad's pretty banged up, but he should be okay".

"Yes, but he isn't making any sense," Julie commented, concerned.

"He was knocked out for some time. He might have a bad concussion," said Scott carefully avoiding telling Julie what my face looked like.

The paramedics arrived and quickly took over. "Okay, Brad, just sit still," one of them said. "We are going to get you on the backboard and put you in a neck brace. Then I'm going to start an IV, and we'll be on our way to the hospital. You took a nasty fall. What happened?"

"I wish I knew," I replied. "All I remember is leaving the house this morning to go on a bike ride. Yeah, that's what I was doing—I was on a bike ride."

"Don't worry about the blood going into your ears. We'll get that bleeding stopped," the attendant promised.

It seemed to take forever to get to the hospital. We finally arrived, and they quickly wheeled me into one of the emergency rooms. A doctor soon entered and asked, "What happened out there, Mr. Henry?"

"To tell you the truth, Doc, I really don't know," I answered. "I remember leaving the house this morning, and then the next thing I remember is waking up on the road."

"What hurts the most?" the doctor quizzed.

"Actually, my hands and wrists hurt the most," I responded.

"You didn't get a scratch on your helmet because your face took the brunt of the fall. You took quite a shock to your spinal column, though. Either that shock is causing your pain, or reaching out to brace your fall is causing it. Since you can't remember what happened, we will have to see what the X-rays show.

"In the meantime, we are going to need to suture some areas of your face that suffered some deep cuts. It looks like a small rock from the road went completely through your upper lip. We will need to suture both outside and inside for this one. You have also cracked some teeth, so you'll need to see your dentist."

Then Julie and the boys walked around the curtain.

"Wow, is it good to see you guys!" I exclaimed. "They are going to patch me up, and then we can go home."

"The ambulance service left your bike at a convenience store near the site of the accident," Julie remarked. "We can get it on the way home, if you like."

"Okay, that sounds good to me. Let's go home. I don't know about Lake Placid again this year, though. That may be a thing of the past

It took about an hour for me to get the X-rays and necessary sutures, but at last I was on my way home. For the next week, I didn't do much. But amazingly, I healed rather quickly. About five days after the accident, I went out for a run. However, it was too soon. The concussion had been a severe one, and it would take me a couple of weeks to get my legs back.

In two weeks, I went out for another bike ride. As I drew near the point where I had crashed, I started to grow nervous. I thought, *I have to do this. If I quit now, I will never get back on my bike.* To say I went slowly through the intersection would be an understatement, but I made it through and continued the ride. I only went about forty miles that day, and I realized that the event at Lake Placid was too close and that I needed to enroll in an Ironman in September.

After looking at the list and talking with my coach, we decided on two races. I would do a Half Ironman in Honu, Hawaii, in two weeks and then the Ironman Wisconsin in September. When I could not do Lake Placid in July because of the concussion I did a Half Ironman in Hawaii and then one in Wisconsin hoping to qualify for the full ironman in Hawaii in October. Even though it was only a Half Ironman, if I had a great race in Honu, I could still qualify for the world championships. However, I would have to place first or second in my age group. But hey, I'm Brad Henry. What could go wrong?

I flew ten hours before arriving at the Kona Airport on the Big Island of Hawaii. I set my bike up and embarked on a training ride through the lava fields of Kona. The temperature was incredibly hot and smothering. All I could relate this to was the idea that I should have practiced in a pizza oven for six months. Then the wind picked up to forty knots, and it was a challenge to even stay on the bike. By this point, I was wondering if I had made the right choice in races.

The day of the race dawned beautiful and began with an ocean swim in a pristine bay. The race director said that although the water depth was about forty feet, it would look to be only a few feet, as the water was so clear. I was excited not only for the chance to swim in such a beautiful bay but also for the opportunity to view the beautiful fish.

Ten minutes before the race, I applied sunscreen and selected one with a high SPF number. The problem was, however, that the lotion was waterproof. That may not sound like that big of a deal to you, and normally it would not be—unless you get it on your swim goggles, which I did. Mr. Henry meet Mr. Murphy.

Ninety seconds before the gun went off, I put on my goggles and couldn't see two feet in front of me. I couldn't get the lotion off, so I ran over to a spectator and asked to use his shirt to wipe my goggles. He reluctantly said yes, and now I could see maybe twenty yards in front of me instead of the usual one hundred yards. It would be quite the swim. Needless to say, my goggles were so bad I didn't see even one fish. I was lucky to see the next buoy.

I finished the 1.2 mile swim in thirty-nine minutes, on course for a chance to secure a spot. I jumped on my bike and headed into the lava fields for the fifty-six-mile bike ride that snaked up the mountain to Hawi then back to Honu. Three—yes, three miles into the ride—my bike tire started to feel wobbly. *No, this can't be happening to me!* I desperately thought. But sure enough, I had a flat tire. I got off my bike and quickly changed the tire, not sure of how much time I had lost. I refocused my effort and returned to the work at hand.

On the way up to Hawi, I started to pass some of the other competitors and felt as if I was getting my rhythm back. Once I hit Hawi, a small town on the northern part of the island, the next ten miles would be mostly downhill. This was the area where the winds were often in excess of thirty-five miles an hour. Riders had to lean their bikes into the wind to prevent being blown off the road. A number of years earlier at the world championships, an athlete had been blown off his bike and had tumbled into a lava field. He ended up paralyzed. Yes, it's crazy trying to stay healthy.

The bike leg of the competition was fifty-six miles, and I was at mile fifty. Again my tire became wobbly—another flat, only this time the back tire. I changed the tire and then used my pressure canister to inflate it, but it didn't release correctly.

Now I had to wait for aid. Ten minutes passed before I was finally on my way again.

I made it to the transition area, changed into my running shoes, and headed off for the 13.1-mile run. I passed about 380 people on the run, but I had gotten too far behind with my bike split to qualify for the world championships. Enduring a ten-hour flight only to have two flats derail me was pretty discouraging. I prayed that I wouldn't get a flat at the upcoming Ironman Wisconsin.

My training was going really well, and I was healthy for a change. I was going to fly out for the Ironman competition on a Wednesday in order to get in a couple of swims and review the bike course. Making final preparations before the trip, I got my bike tuned and was going to do a quick two-mile trial ride around the neighborhood to make sure everything was in place. It had rained the night before, and when I came to a turn in the road, which was on a downhill slope, gravel was pooled in the middle of the road. As I leaned the bike to make the turn, I hit the gravel and went down hard, sliding across the road. I just lay on the pavement, more mad than anything else. I started to check myself over and discovered some deep cuts that might require stitches.

I got to my doctor's office and waited for him to fix me up. It's not too good when everyone in the office knows you by name! Nonetheless, the doctor wrapped my knee, side, backside, elbow, and shoulder, cautioning me not to get into the water before race day. If I had open wounds and swam in lake water, I could get an infection. All of a sudden, things were not looking too good, and on top of all that, I was stiff and sore.

I made it to Madison, Wisconsin, and was able to swim one day before the race. However, every time I put on my wetsuit, my wounds opened up again and I bled. I figured the adrenaline would kick in on race day and I wouldn't feel any pain.

The day before the race featured beautiful weather: eighty degrees and sunny, with no wind. The morning of the race was quite different; the warmest part of the race was actually the swim. Because of a storm system that had come through, the temperature was now fifty-two degrees, with heavy rain and twenty-knot winds for the entire day.

Because of the wind, the swim was really choppy. Additionally, there were so many people in the water that it was hard to pass them. I finally made it out of the water in an hour and twenty-eight minutes, which was very slow time. I jumped on my bike, and within the first five miles, I was freezing. *Only 107 more miles to go! How am I going to do this?* The wind, the rain, and the cold only got worse.

In an Ironman, you usually don't have to go to the bathroom since you sweat the liquid you take in. After the tenth time of going to the bathroom, I should have realized that I was having problems. Around mile seventy, I was becoming hypothermic. It took me seven hours and twenty-three minutes to finish the bike leg, instead of the usual six hours. I could hardly walk when I got off my bike, and someone had to help me out of my wet clothes. I couldn't stop shivering.

I then got something to eat, put on my running clothes, and went out to begin the run course. I remember thinking, Okay, *I can hardly keep my teeth from chattering, and I am having trouble concentrating, so I better stop and fight this fight another day.* I went back inside and gave my race chip to a race official. My race was over—a DNF (Did Not Finish). I did not feel too bad about it that day because I was in such bad shape. But afterwards, that DNF was a bitter pill to swallow. I couldn't help thinking I could have walked most of the marathon and still finished in time. I should have at least tried.

The defeat at Wisconsin did something to me, both physically and mentally. Physically, for the past five years, I have had a hard time getting warm. Everyone else will be wearing shorts, but I will have a sweatshirt on. But mentally, the effect was even worse. I felt like a failure. Many other athletes had been cold and in pain, but they finished.

Ever since Wisconsin, it has been hard for me to find the motivation to train. I think I needed a break both physically and mentally, but I also needed to get back to training. The less I trained, however, the less I wanted to train, until I eventually ended up in a downward spiral of depression and a host of other illnesses.

I have had a very hard time with this particular failure because *I quit.* If you let failure get to you and refuse to try again, it can start a vicious cycle of doubt, depression, poor self-esteem and a host of other problems. But if you get back up and try again, failure can be a great learning experience. Just keep on keeping on!

Remember, pain lasts for a moment, but quitting lasts forever.

Beginning of the End

Over the years, I became great friends with the folks at Joe Gibbs Racing and saw what running a company dedicated to the Lord looked like. For the last race of the year, the head of marketing at the company asked me if I would like to go to Homestead, Florida, with them. I could invite some clients to meet us there, and we could all play golf at Doral Country Club. I said yes before he was finished asking.

I didn't think much about how we were going to get there; I was just happy to be going. A few days before the trip, I got a call to meet at the regional airport at 7:00 a.m. Since I love airports, I arrived early, at 5:30 a.m.

The five of us making the trip walked out onto the ramp, and there it was—the best-looking Lear jet I had ever seen. As we boarded the plane, the pilot and copilot greeted us and motioned for us to turn to the right and find our seats. I suddenly remembered my conversation in the cockpit with the chief pilot Bill twenty-five years earlier. Now I was turning right instead of left, heading for the interior of the plane instead of the cockpit. As I turned right, a sickening feeling in my stomach arose, and a longing to turn left swept over me. Memories flooded into my mind, emphasizing all that I had missed.

It was still dark in the aircraft as I sat down and buckled my seatbelt. The jet engines were started, but all I could do was sit and watch the activity in the cockpit. We taxied out to the active runway and then "into position and hold." I watched the pilot push the levers to full throttle, and the past came rushing back. It was awesome how quickly the Lear reached take-off speed, and we were soon soaring into the heavens. It was all I could do to take in the moment without crying. I had to force myself to think of other things in my life, or I most likely would have broken down. I asked JD Gibbs, the president of Joe Gibbs Racing, if I could go up front and talk with the crew. He knew my past and gave me a thumbs-up.

I spent the entire flight in the jump seat, speaking with the crew. It was if I had been given a second chance. We talked flying the entire way to Miami. At forty thousand feet, we flew over Cape Canaveral, and the view was spectacular. Only an hour after leaving the Charlotte area, we started our descent into Miami. Out came the prelanding checklist, communication with the ground started up, and I felt like I was home again. As we lined up for the runway, the power levers came back, flaps were lowered in increments, and I was taken back in time twenty-five years earlier to my stint at the airline.

As we taxied to the terminal, I realized that God had just given me a gift. He had allowed me to see my past, the present, and the future all in a moment's time. I realized that I might not have gotten saved if flying had not been taken from me. I most probably would have continued living a lifestyle of self-indulgence and self-gratification. I saw that all the setbacks in my life had only pointed the way to *this* day. This day was more about witnessing to others about what really lasts. I had a great deal of fun on that flight, but I realized once and for all that I was content being God's disciple instead of a pilot (at least that is what I tried to convince myself—more on that later). I thought that flying was finally out of my system.

I firmly believe that my flying career was taken away for a reason. I believe that the greatest reason was for me to be saved. God allowed me to have the gift of flying for a time, and then it was taken so that I could go on to the next trial. You see, trials are what build our faith, not great moments of prosperity. Perseverance develops by overcoming setbacks, suffering through pain, enduring through consequences, and relying upon the *only* one who can help—Jesus.

A few years later, the sport of NASCAR took a downturn in licensing. My partner Reed and I had to make a tough decision: we could afford to pay only one of us. It was a very difficult time for both of us. I would like to say that I boldly stepped out in faith to do full-time ministry, but I did not. God used circumstances to bring me to this decision. Reed, who is awesome in business and finances, took over the business, and I stepped into what God had first called me to do on that cold, snowy night in Pennsylvania on November 29, 1992. God saved *me*, a wretched man, a sinner who could never be saved on his own merit. I needed help, I needed forgiveness, I needed a Savior; and he gave all that to me. Even though I didn't deserve it, God came into my life, forgave all my sins, and made me a new creation in him. But the question in life always comes down to "God, what do you want me to do?"

Full-Time Ministry

Being ADD and dyslexic is a challenge when it comes to writing and comprehension. Keeping any thought focused and moving forward is almost impossible. I know, now you are thinking, *How did you write this book?* I have come to the realization that God desires to receive the glory in our lives because He is the one who has created us and given us our talents.

For many years in school, I was called dumb. I couldn't comprehend what I was reading, so I had a hard time learning. The problem with being told you are dumb is this: even if you are not, you start to believe what other people say about you.

A couple of years after I had been saved, I began to read the Bible in the morning and write down a couple of thoughts about what I had read. I first started sending these thoughts to three people. The thoughts that the Lord gave me gradually took on the look of a daily devotional. Now, I am not a deep thinker, but God has allowed me to experience the pains and trials of life. I believe one of the main reasons he did was so that I could have compassion for others and show it in my writing. For some reason, people seem to relate to my simple stories that give God the glory.

Some fifteen years later, the devotional now reaches approximately ten thousand people daily. I know that it is only the Lord who enables me to write. I still have a hard time with comprehension and putting my thoughts together, except when it comes to talking about Jesus. Yes, Jesus is God. Jesus is powerful and transforming. All I have to do is to look back at my life before November 29, 1992, and then look at the change he has made in me. It is nothing I have done; it is only God working in and through me.

You may be in a similar set of circumstances. You may have been told you are dumb and will never amount to anything. Listen very carefully: God doesn't make mistakes! You are wonderfully made by the creator of the universe. The same creator who breathes out stars, holds back the oceans, and shines brighter than the sun is the one who made you. Yet he is also the one who humbled himself to the point of death on a cross so you could feel his love and understand his power in you.

God desires to do miracles in each one of us, if we will allow him. However, we are often too busy or too prideful to allow him access to every part of our lives. This life is not about how many times we get knocked down; it is about how many times we get back up. So many times it is easy to accept defeat in the moment of extreme trial. But as most of you probably know, giving up in defeat leaves a very bitter aftertaste. Remember what Lance Armstrong, the multiple Tour de France winner, said: "Pain lasts for a moment, but quitting lasts forever."

Trials build faith. The longer and more intense the trial, the stronger the faith created. There is no shortcut to this process. In a world where we have sped up everything from the way we download documents to the way we order a meal at a fast-food restaurant, we demand quick solutions. When we carry this attitude into our relationship with God, we find that we expect an answer to when our particular trial will end. But God is the only one who knows, and we are not privy to that information. We are privy, however, to the one who holds that information.

God will not let you go through a trial for nothing. There is always a perfect purpose in any trial, even the one you are facing as you read these words right now. If God can use a C minus—okay, D—student to write for him, he can use you in ways you could never imagine.

The title of this chapter is "Full-Time Ministry." Let me clarify what this means for you. All of us—yes, all of us—are in full-time ministry. If we have been saved by the grace of God, then we have a story to tell. Isn't it amazing that God uses *us* to reach other people? He doesn't call down from heaven in an audible voice to save us, even though he could. He uses someone else as a mouthpiece for him. The question now is, if you have been saved from hell and have a secure place in heaven, not because of what you did but because of what Jesus did in you, why not talk about it?

There are a number of reasons we don't do this, but the main one is we don't understand what we have in Jesus. We forget about our original sentence of death as a fallen man or woman in this world. If we came up with a cure for cancer, we would go to every talk show and newspaper in the world to share our wonderful discovery. So why don't we have that same enthusiasm to share what we have discovered about enjoying eternal life with Jesus in heaven? If I had cancer and was cured today, guess what? I would still die. But if I have eternal life in Jesus, I will live with him forever in heaven, even though my body dies.

Jesus brought Lazarus back from the dead, but Lazarus still died later in his life. Everyone in the Bible who died but was brought back to life eventually died again. So why did Jesus bother to even bring them back to life? First of all, Jesus knows everyone's heart. He knows that miracles in themselves will not bring people to him. If someone has a hardened heart, no miracle will penetrate it. But Jesus raised people from the dead so that we could see his power over death.

There are many prophets and religions in this world. But Christianity is the only religion in which the creator died and was brought back to life. Every other person who started a religion is still dead. You see, when Jesus died on the cross, there was no power. In fact, all his disciples were afraid and abandoned him. But Jesus demonstrated his power when he conquered death and rose again. Now those who believe in him will never die but will inherit eternal life.

If we are saved, we will enjoy eternal life in heaven because of the one who now lives in us. After we have been saved by the power of Jesus, we are then asked to go and share with others what God has done in our lives. People can always refute the Bible, but they cannot refute what God has done in another person's life.

Phoenix I took a class called Evangelism Explosion, which was led by my friend Pastor Bill. This was a thirteen-week class that provided all the tools needed to lead others to a saving knowledge of Jesus. One day while I was waiting to catch a flight from Los Angeles to Phoenix, I had the opportunity to use what I had learned. Southwest Airlines had a flight leaving for Phoenix every half hour, so I just showed up at three that afternoon, expecting to catch the three-thirty flight. "I'm sorry, Mr. Henry," I was informed, "but we are booked solid until 6:30 p.m."

"That's okay," I answered. "If anyone cancels, please let me know, but go ahead and put me on the six-thirty flight. I'll be over in the food court."

I went to the food court and spotted only one empty table. As I was getting ready to sit down, another person sat down at the same time. I said, "Oh, excuse me," and I prepared to go sit near the gate.

But the man, who had muscles and tattoos galore, replied, "Hey, no problem. You can sit here too."

"Great, thank you."

All of a sudden, a shadow formed over the table. A man who was seven feet tall and weighed 550 pounds was standing there, ready to join his friend at the table. There were three men in all, and they all sat down and shared with me, yes that they were professional wrestlers flying back to the East Coast. For the next thirty minutes, they shared the vilest things about what they had done in the past couple of weeks. Their conquests with women and hard drinking actually brought back memories of the lifestyle God had saved me from. All the time, I was praying for an opening to share Jesus with them.

Then, in answer to my prayer, one of the wrestlers asked, "So what do you do?"

Without any thought on my part, the words came tumbling out: "I am an evangelist."

One of the wrestlers quickly replied, "You are a priest and we've been talking like this? Sorry."

I told them I was no priest but had two questions for them. They all gave their permission for me to ask the questions. First I asked, "Have you come to the place in your spiritual life where if you were to die today, you would go to heaven?" Most of them answered yes. I followed that up by asking, "If you were to die tonight and stand before God and he were to say, 'Why should I let you into my heaven?' what would you say?"

One of the guys exclaimed, "My mom just asked me that question!"

Then I asked the men if I could share the good news of the gospel with them. They all consented. For the next twenty minutes, I told them all about my past. I shared how God had saved me, not based on anything I had done, but by what he had done for me. I told them that we are all sinners in need of a savior, that heaven is a free gift that cannot be earned and certainly something we do not deserve—especially because of the lifestyle so many of us are living.

Then I told them the only way to inherit eternal life is to call upon the name of the Lord. I explained that we have to ask Jesus to forgive our sins, and we have to acknowledge who he is and what he has done. Then, I concluded, we need to ask him to come into our lives.

I asked them if all this made sense, and each one of the men said yes. Then I asked, "Would you like to receive the gift of eternal life right now?" Again they all said yes. I told them we could pray right then and there; I would say a prayer one sentence at a time, and they could repeat it after me.

There were more than two hundred people in the food court that day, and these men at my table really stood out. But they all unashamedly pushed their chairs back from the table, bowed their heads, and waited for me to lead them. I prayed the prayer, and they prayed after me. That evening in a food court in Los Angeles, California, God saved three hardcore wrestlers. He used a man who had been just as immoral as they were. God's destiny was for me to be there at that time so these men could hear his word and receive him.

We hugged each other, and then they left for their gate as I walked to the Southwest counter. They had one seat left on the 5:30 flight, so on I went and off I flew to Phoenix.

As David Ring says, God doesn't need our ability; He desires our availability. God set up my appointment with the wrestlers that day, just as he has with all my other appointments since. Each time it has been God's words spoken through me that lead others to repentance. The problem comes when we arrogantly think that we play a part in any event God does. God is King, Creator, and Redeemer, and we are not. God has had me sit beside many people on airplanes and in other places who have been ready to hear and thus received his word. But it was God who did it—not me.

God has you right where he wants you in order to meet people that others can't reach. All of us have a circle of influence. The question right now is, are you using this influence to further your kingdom or to further God's?

Don't be too hard on yourself if you felt a blow to your heart with that question. In a world driven by performance and material things, we can easily get distracted from the real goal. Don't live in past failures or mistakes. Realize that the Lord can give you a clean slate every day when you ask for it with a repentant heart. Then live your life as unto the Lord. Don't let anyone tell you how that will look. Only the Lord has that right.

Life's Storms

Julie and I are blessed to have two boys. Bryce is now fifteen, and Chase is twelve. People have asked me what the difference is in my life since becoming a Christian. The falsehood that some new believers embrace is that things are supposed to get better in their lives after they become Christians. I have learned, however, that strong storms still come, but now I don't have to handle them alone. Storms come for a purpose, so now when I face one, I ask God to help me through it, whereas before I had no one to help me.

One of the biggest storms Julie and I have faced is that our son Chase is autistic. Chase seemed quite normal until he was about eighteen months of age. He hit all the developmental milestones that every other child makes. He crawled at nine months, walked just after his first birthday, and was soon babbling up a storm. Then, around eighteen months old, he stopped speaking and seemed to withdraw from us. We took him to our family doctor and voiced our concerns about Chase's lack of speech, but the doctor shrugged it off and said, "He'll talk when he talks. Don't worry about it." Others tried to encourage us, saying, "Einstein didn't speak until he was four years old." I'm not sure if that was the case, but it eased our minds.

Have you ever known what you were seeing but just didn't want to believe it? Well, that's what happened with Julie and me. But finally, when Chase was four years old, we took him to be tested. He went through a battery of tests, but Julie and I already knew the diagnosis. The day came when we were to get the full report, and we walked into a conference room at the testing facility. We sat on one side of a table, and three doctors sat on the other. One doctor spoke quite casually and seemed to be quite happy with himself for the diagnosis he had discovered. As he was speaking, I wrote on a piece of paper the word *jackass* and slipped it to Julie. Being a Christian enables us to have strength in the storm, but we don't always get everything right. This was our child, our boy, this doctor was talking about. There was no emotion—just a diagnosis and good-bye—and I reacted to that.

It has been eight years since that diagnosis, and I would have to say that Chase has taught Julie and me more than we have taught him. If you know any autistic children, you know they often fixate on unusual things. Chase, for example, could be paid by McDonald's for how many times he has mentioned their name.

Chase also loves the cups at Bojangles', a fast-food restaurant in the South that features chicken and biscuits. We often go through the drive-through and ask for water just so Chase can get a cup. When Chase got on the Bojangles' website, he saw that their coffee cups looked quite attractive, so he and I went into Bojangles' and asked to buy a cup, but no coffee. After the puzzled look I got, I explained what we were doing. The man was so compassionate that he gave us not only five coffee cups but also the lids to match. I had one happy boy. Okay, I was quite happy too.

Chase also has to watch all the clocks in the house turn to the fifty-minute mark: 7:50, 8:50, and on and on all day. At forty-nine minutes past the hour, you'd better not get in his way, or he will knock you down as he runs through the house. You see, he has only one minute to look at all the clocks before they turn to the next minute.

I made the mistake of introducing Chase to the Three Stooges. Well, let's just say that although Chase loves the Three Stooges, Julie definitely does not love the effect they have on him! For instance, Chase and I were standing in line at the post office one day, an experience that can be pretty exasperating for anyone. People are looking at their watches, huffing and puffing, and God forbid if anyone wants to carry on a personal conversation with the clerk. On that particular day as Chase and I waited in line, he suddenly exclaimed in a loud voice, “Nuk, nuk, nuk! Wuuup, wuuup, wuuup! Dad, do you want to watch the Three Stooges tonight?” Well, the man who was huffing and puffing started to laugh, and so did everyone else in line. I, of course, was horrified. *Dad, can we watch the Three Stooges? Yes, Chase, yes. No problem—just be a little quieter!*

But that is Chase, and strangely, if he woke up tomorrow morning and was cured, we would miss the Chase we have now. Chase has taught us that almost everything in life is not really that important. He has taught us to have fun in the moment. He is not too worried about anything in the future. He has a tender heart that shows the compassion and trust that Jesus talked, the kind of heart that we need to come to Jesus with the faith of a child.

Chase’s brother, Bryce, has also learned a lot from Chase. He has learned that life is not always fair. He has learned that we can’t do everything as a family that we would like to do, but the good news is that we have a family. I can see Jesus speaking through Chase, and I can see him using Chase to refine us as a family.

Many Christians go to church every Sunday, serve and teach in Sunday school, give their 10 ten percent or more, and then expect automatic favor from the Lord. However, doing all these things does not guarantee that storms will not come. The only reason we should do good things is out of gratitude for what God has already done for us.

Many people get burned out serving because they are constantly trying to get in better with God. If you are saved, you will never get in better with God. He loves you just as you are. The problem is, the world views us based on our performance, and God doesn’t. That is a huge paradigm shift. That is why we need to constantly be in the Word (Bible), pray, and maintain fellowship with other believers.

We need to know who God is. If we can ever understand this, we will stop the performance game and just love God and others, not expecting anything in return. I think we all need to be a little more like Chase. Julie and I are so thankful that God saw fit to give us this unique, precious boy, as well as his big brother.

Oh, oh, it's forty-nine past the hour—better look out!

Natural Stick

After I left HLG Licensing, the company I formed with my friend Reed, my financial future was horrific. Living by faith and on donations from others is certainly a challenge. But it is also a miracle. Julie and I might get down to only seven dollars in our checking account and need four hundred dollars, and then a check will come just in time—not early, but just in time. I am a person who hates to wait, but God has taught me so much in this process. You see, God knows my heart and how to change it.

Once I was in a big hurry as I pulled up to the drive-through at McDonald's and ordered a Coke, fries, and a hamburger. They told me the amount and directed me to pull around to the first window to pay, which I did. While I was driving down the road, a car suddenly pulled out in front of me. I extended my hand in order to keep my Coke from spilling, only to realize that I didn't have a Coke. I didn't have any food either. I had paid for my food and then driven right past the people waiting to give me my dinner. Sheepishly I drove back, and almost every employee wanted to see the guy who had left without picking up his food. Yes, I did get my food.

God knew that I needed to develop patience. Living from ministry funds, however, became so tough that I began to wonder if I could or should do something else to generate income. I never expected what happened next.

"Brad, did you see that article on MSN today about the FAA allowing new medications that were prohibited in the past?" asked my wife.

"No Julie, I didn't see it. What section is it in?"

"Actually, it's on the front page of MSN."

"Okay, let's see what this is about," I responded. As I read the story, I just about dropped my laptop. The one medication responsible for the loss of my pilot's license was now approved by the FAA.

"Are you kidding me?" I exclaimed in disbelief. "After all these years, now I can fly again?" I was incredulous.

Have you ever been afraid to dream? Well, I have. I had stuffed my feelings about my desire to fly for so long that to now have the opportunity to get back into an airplane was beyond my comprehension.

But is God calling me to ministry? Is it selfish to follow a dream or to fulfill a passion? Is this passion God's desire for my life, or is it a trick of the evil one to distract me from what God has called me to do? All these thoughts ran through my brain in about three seconds. Being ADD and dyslexic has its good and bad points. I can think at light speed, which is good, but the problem comes in harnessing those thoughts into a well-organized plan.

Across the street from us lived a check airman for a major airline. A Check Airman flys in the airplane with you and signs you off if you are qualified to fly after all your ground school and simulator training has been accomplished. I immediately called her. "Jill, I have a huge favor to ask," I began. "Is there any chance that you could get me into the 737 flight simulator? I am thinking about getting my medical back."

"Well, ever since 9/11, it's been more difficult to get outside people into the sim, but I'll check for you," she helpfully offered.

"Jill, that would be awesome! No worries if you can't, but let's just say that would be my birthday present till the day I die. But no pressure, okay?"

Hanging up the phone, I turned to my wife. “Julie, you’ll never believe what just happened. Jill may be able to get me into the sim!”

“Don’t get your hopes up just yet,” Julie cautioned.

“I know, Julie, but to fly again would be just unbelievable.”

I didn’t hear from Jill for a couple of weeks and assumed the idea was dead in the water. Then I received an e-mail from her. “Brad,” it read, “not only are you good to go, but so are Bryce, Chase, and Julie. We have the sim for four hours. Would that be okay with you?”

What? Okay with me? You don’t know what this means Jill! Wow, wow, wow!

The day of the sim ride came, and I was like a kid on Christmas morning. My family and I arrived at the training facility, and after we signed in, Jill showed us around. We saw where the flight attendants trained and where all the simulators were located. It was an impressive sight. Each simulator sat atop ten-foot hydraulic stilts that could rock back and forth to simulate actual flight conditions. The simulators were so good that prospective pilots could do all their training in them. Their first flight in an actual airplane was with passengers on board.

“Okay, Brad,” said Jill, “jump into the left seat and let’s see what you remember. We are at a weight of 121,500 pounds for takeoff. The weather is clear, temperature sixty-one degrees, winds out of the west at five. We will take off on runway 27 right.”

I secured my five-point harness and made sure Julie and the boys had their seatbelts on. Since you are at least ten feet off the ground in the simulator, the motion can throw you out of your seat, so it’s important to be strapped in.

“Okay, Brad, V1 is 136 knots, Vr rotate, 139 knots, V2 142 knots. At 1800 feet, initial flaps up, and maintain runway heading. Cleared for takeoff runway 27 right.”

“Roger, Worldair cleared for takeoff runway 27 right.”

As I grabbed the throttles, the twenty-seven years since I had last flown felt like only a day. I inched the throttles up and pressed the TOGA (Take Off and Go Around) switch. The throttles automatically advanced to take-off power.

"120 knots, 136 knots, V1, Vr, rotate," came Jill's call out's.

"Roger. Positive rate, gear up," I responded.

"Gear up; maintain 180 knots. Turn left heading 180 and maintain 3,000 feet."

"Roger. Worldair turning left 180; climb and maintain 3,000 feet."

"Okay, Brad, would you like to try an approach into 27 right?" Jill asked. "We'll make a ceiling of 500 feet and visibility of one mile."

"Sounds great, Jill. Let's try it."

"I have the glide slope and localizer set. Speeds are 180 knots till the outer marker, then gear down. Flaps 15, speed 150 knots. Turn left heading 330 to intercept the localizer for runway 27 right."

"Jill, the localizer is coming in, and the glide slope is active. One point five dots above, glide slope active, gear down."

"Julie, do you see this?" called Jill.

"I can't see anything," said Julie.

"Your husband is flying better than half the people I get in here who are flying full-time. Some pilots have a special gift, and I call them a 'natural stick.' Your husband, Julie, is a natural stick."

"Okay, 1,000 feet above the ground, 700, 500. I can see the runway lights, full flaps," I commented.

The voice on the terrain system called out in a deep, hollow voice, "Fifty feet, thirty feet, ten feet," and then I could feel the wheels touch down. I grabbed the reverse thrust, pulled up and back, applied the brakes, and taxied to the gate. That night we did the river approach at Washington National and another low approach into a two-hundred-foot ceiling and visibility of half a mile.

"Too bad you couldn't get your medical back, Brad," said Jill. "It wouldn't take you any time at all to get back to speed. Actually, you could fly the airplane now; you would just need a refresher course on the regulations and emergency procedures."

I was very grateful to Jill for giving me a chance to fly again, even if only in a simulator. "I can't thank you enough," I gushed. "This was a dream come true. I will never forget this."

"Any time, Brad. Julie, I hope you and the boys had a good time too," Jill answered.

"Just seeing the look on Brad's face was all I needed," my wife responded. "You made his dream come true."

My family and I got back into our car, and I could not stop talking about flying. That next week I signed up with the FAA doctor to try to get my medical back. The day of the appointment finally came, and after the examination, the doctor said, "Okay, Brad, your health is great, but the medication you are on needs to be cleared by the FAA." *The medication you are on.* The last time I had heard those words from a doctor had been twenty-seven years ago, but the horrible memory seemed like only yesterday.

"Brad, did you hear me?" asked the doctor.

"I'm sorry, Doc. My mind was somewhere else," I sputtered.

"I said that we need to send a copy of the medicine you are taking to Oklahoma City. It may take some time, but hopefully, you can get back into the air soon."

"Thanks, Doc. I'll let you know what I hear," I replied.

Even though the new medication was approved you still had to get confirmation from your doctor of how long you had been on the medication, if you suffered any adverse affects and then have your doctor sign a statement that everything was correct.

Lord, please help me get my medical back. It would seem like a cruel joke to let me get this excited again and then not let it happen. Ten days later, I got the letter I had been waiting for from the FAA. But it produced more doubt than hope. It said that I needed to wait ninety days before I could begin the process.

Was I missing something? I had certainly felt that the Lord was giving me a green light, but now I was somewhat confused. I spoke to my regular doctor about my predicament, and he suggested something that could have come only from God. "Brad," said Dr. Hall, "you are taking such a small dosage of the medication. Why don't you try getting off it completely?"

don't know why that thought had never entered my mind. Many times, I think, we take medication prescribed by a doctor and then just keep taking it even after we are okay. Then we have to start taking other medications to combat the side effects from the first one.

I went home and spoke to Julie, and though she urged me to be cautious, she thought the doctor's suggestion was definitely worth a try. Within a four-week period, I was totally off the medication and suffering no side effects.

I called the FAA to inform them that I was now off the medication that had caused the original problem on my medical. All they needed now was a letter from my doctor stating that I was off the medication and not having any adverse issues. Could it really be that easy after all these years? Could it be that after twenty-seven years God would give me back my dream? Now I would wait.

The FAA had just increased the age limits for commercial pilots from sixty to sixty-five years. I had ten years left. Was it worth it? I spoke with Julie, and she realized I had dreams that still needed to be lived. We die in this life when we stop dreaming and start thinking we are too old to do something. Without dreams we lose our God-given purpose in life. This dream of flying again was a dream I did not want to wake up from.

Julie and I had many discussions about the possibility of my taking to the air again. "Most of the major airlines look for two thousand hours of flight time and two hundred hours of multiengine to hire someone," I explained to my wife. "I have over five thousand hours of flight time and over two thousand hours of multiengine."

"But, Brad, you haven't flown in twenty-seven years. How can you get over that hurdle?" asked Julie.

"Great question. I just need to get in front of people and explain my story. Hopefully, they will understand and give me a chance," I replied.

"But remember, you need to get your medical back first," Julie cautioned.

"I know this is going to be a tough wait," I remarked. "I didn't have a problem when I thought the dream was out of reach. It's almost like being rescued after twenty-seven years on an island. I was afraid to ever hope for it. But now that I've opened my heart again to the possibility, the wound will be even deeper if it does not happen."

"Well, that's true," Julie thoughtfully commented, "but you tell me all the time to chase after my dreams."

"I know," I laughed. "It's so much easier to give advice to others than to take your own advice!"

Changing topics, Julie reminded me of the evening's plans. "Do you remember that tonight we have our small group meeting at the Von Olhausens' home?"

"Oh, that's right," I responded.

"Kim is going to share her testimony. I think her first husband died of some illness. I heard we better bring some tissues. Let's leave the house at six thirty."

"Sounds great, Julie."

Redemption

Have you ever been caught off guard? Have you ever expected God to do one thing and then all of sudden—bam! Well, you know what I mean.

We arrived at the Von Olhausens' house and were warmly greeted. "Hey, Brad and Julie, great to see you tonight. The kids are playing upstairs, so Bryce and Chase should have a good time. Help yourselves to some coffee and brownies, and we'll start in a few minutes."

Every other Wednesday night, about six couples from our church met for fellowship. This night Kim was going to share her testimony.

"Some of you know bits and pieces of my testimony," began Kim. "But hopefully, I can tie it all together tonight." Kim then shared that she had lived a pretty normal life until she got married. Her husband, Chad, was a very jovial guy, and everyone liked him. One day at work, Chad severely injured two discs in his back. He had to start taking painkillers just to make it through the day. Six months later, he was addicted to them. He soon needed more and more of the drug to get through the day, so he started going to different doctors to get more medication. However, he managed to hide the severity of his addiction.

One weekend some friends asked Kim and Chad to go camping with them. Chad declined because of his bad back, but he encouraged Kim to go without him. Leaving the house that day, Kim had no way of knowing that it would be the last time she saw her husband alive.

While Kim was gone, Chad robbed a pharmacy to get the medication. He was now a wanted criminal. Kim spoke to Chad on the phone and told him they could work it out, but Chad would not listen. One night in a motel, Chad could no longer bear the shame of what he had done and took his life. What Kim said next would forever change the way I saw God. “Chad could not accept God’s grace,” she sadly said.

That described me perfectly. I had grown up in a performance-based home. Even though I was saved, I viewed my heavenly father like my earthly father. When we got home that evening, I shared with Julie everything that was going through my mind.

Ephesians 2:8–9 says, “For it is by grace you have been saved, through faith—and this is not from yourselves, it is the gift of God—not by works, so that no one can boast.” I knew this verse in my head, but not in my heart. Even though I knew God’s love was real and present in my life, I always felt in the back of my mind that I had to do more. I could never rest in God because I had a warped view of God’s love. But listening to Kim that night, I finally got it. Sadly, it took the tragedy and pain of another family for me to be able to accept God’s grace.

A friend of mine who helps train National Hockey League players told me that God’s grace is playing the game of life without a scoreboard. In life things can get tense when we keep score. *If I do this for you, then you have to do that for me.* With that mind-set, we do things out of duty instead of love.

How would you live your life if no one kept score? I have listed some thoughts on how you can love without asking or expecting anything in return:

1. Since God can’t love you anymore today than he did yesterday or than he will tomorrow, why do you do the things you do? Is it performance based?

2. When God doesn't immediately answer your prayers, that doesn't mean that he doesn't hear you. In order to learn patience, you need to *slow down* and listen.
3. God's grace *is* sufficient. Forget performance. Do things out of a thankful heart for what God has done for you—period! Since you cannot gain better standing with God, then performance-based thinking is the opposite of grace. If grace has to be earned, then Christ died for nothing.
4. Life is full of ups and downs. The ups and downs don't have anything to do with your performance in life.
5. God at times may let you wander. You can't be going 110 percent all the time. If you try to do that, God will sometimes get your attention by allowing you to wander in the desert for a while.

For some strange reason, in the two weeks after my doctor's appointment, I kept seeing this verse: "Be still, and know that I am God (Psalm 46:10). *But, God, you know me—I can't sit still*, I silently protested. *Plus, how will I get things done if I have to be still?* I tried to push the verse from my mind, but it kept cropping up again and again.

Finally, after much deliberation, I realized that "be still" didn't mean I was to stop working or trying to do my best. God was telling me to let my heart be still. I had experienced chaos all my life, and my tendency was to try to bring everything under my control. Ever been there? But God was telling me to just trust him.

For the first time in my life, I started to realize that the devil was causing me to live a lie. I was believing what Satan told me instead of living for the glory and hope of what Jesus had given me. I had suffered so many setbacks that I felt God would never deliver something good to me now after all these years. But regardless of the outcome concerning my flying, I was ready to trust God. Whether I ever flew again or not, I was going to trust him.

God’s promises and his love for you and me are perfect. If we are not still in our hearts, then we will mess up what God has planned. For once in my life, I was going to be still. I finally began to realize that my failures were not a reflection of my being a failure. God loved me—that was enough.

Will I Get to Fly Again?

"Depression!"

"Yes, Prince."

"I banished Panic for his failure to stop Brad in his quest to get to know J. Bring Panic back. I have one more task for him. If he can do this, he may yet redeem himself with me."

"Yes, Prince."

....

I had spoken to my doctor and since I was on such a low dosage of medication we discussed getting off the medication completely. Then there would not be any reason I could not get a First Class Medical. So the doctor gave me a plan and for almost 3 months I had no side effects. I was ecstatic. So I sent another letter to the FAA stating that I was completely off the medication.

The long-awaited letter from the FAA Medical Certification Branch finally arrived. This was it. Twenty-seven years of my life was hanging in the balance in this document. Part of me started to think, *Why even open it? It'll just say, "You will never fly again."* But this time was different, as I've explained. This time I was going to trust in God instead of my own wisdom.

This is what the letter said:

Dear Mr. Henry,

We still need more information from your family doctor concerning the medication that you have stopped taking. You need to be completely off the medication for 90 days without any issues before we can consider issuing you a medical certificate. After ninety days, you can reapply for reinstatement for your Class 1 Medical.

This was awesome news. I had been waiting so long for this letter that it was already eighty-six days since I had last taken the medication, and I had not experienced any side effects.

"Julie, I spoke to the folks in Jacksonville, and I can take a seven-day jet-training session to get me back up to speed. They said that if I pass this qualification, I'll have an interview for first officer in two months!"

"Aren't you jumping ahead a bit?" Julie answered. "You don't even have your medical back yet."

"I know, Julie, but if I wait till I get my medical, then I'll have lost precious time. Plus, the airlines, after a ten-year layoff, are now hiring. I can't believe that I'm finally having good luck!" I exclaimed excitedly.

"I know, Brad. You have wanted this for so long, and I can hardly believe it myself."

"I'm so excited," I chattered. "To tell you the truth, to fly again would be something beyond words. It's all I can think about. What a gift from the Lord!"

Shifting to more mundane conversation, Julie suggested, "Brad, why don't we all go in the backyard and sit on the porch? It's such a nice evening."

Julie was right; it was a beautiful evening. As I sat on the porch with my family, I looked up into the sky and saw a contrail. A contrail is the line a jet makes in the sky when the hot exhaust from the jet engine hits the cold atmosphere and forms ice crystals.

"Bryce, can you believe it? I could be up there in a couple of months doing that for a living again," I remarked to my son.

"Wow, Dad, that would be so awesome! How high do you think that jet is?"

"Oh, maybe thirty-five thousand feet, Bryce."

Out of nowhere, it hit. I started to feel flushed, and my heart was beating faster than normal. *Oh not, this can't be happening!* I despairingly thought. *There are only four days left until the ninety days are up!* I took my blood pressure, and it was 165/90, much higher than my normal 120/70. Julie, noticing something was amiss, asked me what was wrong.

"Julie, I think this is it," I answered. "I think my flying is done."

"Why don't you take your blood pressure again?" she suggested, trying to soothe me. "It may have gone down."

"It sure doesn't feel like, it but I'll take it," I replied. Well, I took it a second time, and it was now 195/95.

"Julie, I'm going to have to take the medicine," I concluded.

"Are you sure, Brad? Are you sure you need to take it?" Julie asked.

"Yes, Julie, I do. I'm afraid my blood pressure will only go higher, and then I could have some real problems."

As soon as I took the medication, a peaceful feeling swept over me. Of course, it would take thirty minutes before the medication could be effective, so I knew this feeling was not coming from the medicine—it was the presence of Jesus with me. Jesus knew what flying meant to me. He knew that it was more than a part of my life—it was my whole life.

I went downstairs to talk to Julie, to tell her my flying days were over. Still at peace, I told her I understood why God had allowed this to go on for so long. I had harbored resentment over losing my flying career twenty-seven years earlier. I had watched as all my friends made captain at major airlines, and I had been jealous. I had stuffed this feeling deep inside for many years, but God knew I had to deal with it. What better way to finally put the entire issue to bed than by forcing me to face the possibility that I could still do it?

....

"Prince, I did my job, but Brad does not seem to be hurt by this," Panic explained. "If anything, he seems to have a stronger resolve."

"We have some big problems with this Brad Henry. I want him attacked from morning to night, you hear me?" Satan bellowed. "Bring in the other demons who have attacked Brad in the past. This needs to be an all-out effort. He cannot talk about his victory to anyone. We *cannot* lose souls to J that are ours. Do you hear me?" he roared.

"Yes, Prince, we will not let you down," promised the quivering demon.

"Somehow I want Brad to feel defeated," Satan ordered. "If he feels defeated, he will not be any good to anyone. Most of all, prevent him from praying. When we get people tied up in themselves, they don't have time to access the power of prayer to fight us. Then defeat will come, and defeat breeds discouragement and loss of self-esteem. That is a great tool we can use, right, demons? *Right, demons?"*

"Yes, Prince," the demons obediently responded.

"So go out and keep destroying! Keep giving people what they want as long as those wants are self-serving."

"Yes, Prince, we will."

....

I had shared in my daily devotional about my desire to fly again and all that had been happening for the past four months. When the months ended as they did, a number of people wrote and said they had wept when they read the story. They knew my love for flying, and for me to get this close to having it back again seemed absolutely heartbreaking to them. But in the end, God worked flying out of my heart the way you work a splinter out of your hand.

Like me, you may have deep-rooted things in your life that God needs to draw out. Remember, he does it for your own good because he loves you. When I lost the chance to fly again, I could see that it happened because God loved me, not because he had it in for me. As a result, I had peace. Twenty-seven years earlier, I did not have Jesus in my life, but now I did, and he enabled me to see my heart and my desire.

what was my desire? I realized that the thought of flying again had taken over a huge part of my life where Jesus had been firmly ensconced, and it was not a pleasant thought. I acknowledged that flying for countless hours from point A to point B would have been fun, but sharing the good news of salvation and how to live a life for God was what I was called to do. God had given me an incredible gift to write the daily devotional, and that was my calling—not flying.

As I write this chapter, I am saddened beyond belief—not because of the flying, but for another, more tragic reason. You see, Jill, the woman who gave me my life's dream of flying in the simulator, the one who called me a natural stick, was brutally murdered a couple of weeks ago in her house along with her husband. They lived across the street from us and brought us much joy. Not only did I enjoy talking about flying with them, but I also considered them great friends.

The funeral brought back a lot of tough memories. Of course, I was grieving the death of my friends, but every uniformed pilot in attendance brought home the point that I was no longer among that special group of professionals. My old roommate, who is now a captain for an airline, walked in the door, and we hugged. Missing the camaraderie that I had shared with my pilot friends, I felt the old bitterness trying to sneak in. But in the back of my mind, I knew then and I know now what God called me to do.

As I stood in the receiving line at the funeral, one of the relatives of the family asked me if I was Brad Henry. I was a little taken aback that this person knew me, since my friends' families were from Texas and Georgia. But only days earlier, I had written a daily devotional about the tragedy, and somehow it ended up in their hands. I found out who had forwarded the e-mail to them and contacted the person. This is what he wrote to me:

> First, I am deeply saddened by this horrible tragedy. I hurt for this family and for you and your community as well. I know your loss and the emotions felt; I also know it is encouraging that events like this only motivate you and God's people to work harder because we have only a short time here and our mission is the most important thing. So with loss and pain come renewed spirit and clear direction—God's plan is perfect. Still, death and evil are overwhelming. Nothing happens by chance with God, though. We are in three different states, and yet all of us are connected by God's words through a devotional he blessed you with.

At this point in my reading, I paused for a moment, overcome with emotion and divine revelation. *You're still a pilot. Maybe you're not behind the controls in a cabin cockpit, but you're safely flying God's people towards heaven, and that's even more critical than safely landing them at an airport!*

I continued reading:

> Thanks so much, Brad. It is still strange to me, even in our global world and the instant communication it provides, that these types of meeting can occur; but then again, with God . . .
>
> I love you, Brad, and thank God that on this occasion, someone loved me enough to pass along the devotional. They listened and followed the Spirit, and here we are today. May God continue to bless you. I am certain the words he gives you in your devotional will continue to multiply to hundreds of millions in the same way he brought them to me. Keep on keeping on!
>
> Thanks, Brad!

Then, a relative of the murdered victims wrote this to me:

Good morning, Brad,

After a very long weekend of traveling back to Georgia, spending time with family, taking care of yard work, etc., we are back at work this morning and catching up on your devotionals. It was so nice meeting you last week, even though it was through the tragedies of Bobby's aunt's and uncle's deaths. I truly believe that things happen for a reason. Had our coworker and friend not forwarded the devotional you wrote, we probably would have just shaken your hand in the receiving line and never had the opportunity to get to know you better.

It's funny, but before I knew Christ, I would often wonder how people knew when the Lord was speaking to them. Did they actually hear his voice, or did something happen in their life and God spoke to them through someone or something? I now know the answer is *both*. I know that God has spoken (and continues to speak) to so many through your devotional. Even in this time of tragedy in our family, God is speaking to all those around us.

Let me just say, that of all the wonderful pilots I met last week, you are the most memorable. And it's not because of your ability to fly a plane, but rather for the work you do to spread God's Word!

Andrea

God has so graciously shown me that when he takes something important away, he always replaces it with something better. God sent these two people mentioned above at just the right time to let me know I was on course in my life.

Maybe there is something in your life that you have not yet let go of, and it is eating away at you. This thing is robbing you of joy that could easily be yours. Please ask the Lord right now—yes, right now—to take this burden away.

I know you want to become better, but somehow you have become bitter. But remember, Jesus came to give you not only eternal life later but also abundant joy now. No matter what you are holding on to, Jesus can take it and give you something 100 percent better. All you have to do is ask. If you are struggling with someone or something, then please pray this prayer with me:

> Lord, please forgive me for harboring resentment for __________. Lord, you forgave me for all my sins, and now you ask that I forgive others. Lord, I want to forgive, but I have held on to this feeling for so long that I don't know how to give it up. I have become dependent on this worry. Please, right now, Jesus, take this pain from me and give me your joy. Please give me a stronger faith to let go and to live the life and be the person that you intended me to be. Amen.

God's Perfect Plan

In the process of writing this book, I thought of many great ways that it might end. Sometimes I envisioned myself in the cockpit of an airliner, living out my life's dream. Other times I imagined being the CEO of another company and financially set for the rest of my life. Maybe Chase would be healed of his autism, and on and on the imaginings went. But in all these scenarios, the ending was still about me.

It has taken me longer to write the last two chapters of this book than it did to write all the previous chapters. That is because God needed to reveal to me my motives. I have written much about my sinful behavior when I was younger. I did that to show how God can take a sinful, broken life and transform it into something that brings him glory. I wrote about the many awful things I did, certainly not to boast, but to show that the Brad Henry of the past is not the Brad Henry writing this book today. I have been changed by Jesus, not by myself. I will never be the same!

I also wrote about my many personal failures in order to remind myself of God's goodness, to emphasize that *no one* is too far gone for God to save. I wanted to encourage you that just because you have failed in the past doesn't mean that you will fail in the future. I wanted to show you how God is faithful even when you are not. And most of all, I wanted to encourage you to live a life that is full of joy and not despair.

Thinking of all these things, I eventually realized that this book needed to end in a way that was not about me at all. God saved me when I didn't deserve to be saved. He loved me when I didn't deserve to be loved. To finish the book with a great story of my success would be against everything God has brought me through. The only way I can demonstrate God's love to others is to let go of my pride and arrogance and make the ending of my story all about him.

My greatest glory should be in serving the King, the Creator of the heavens and the earth, the God who saves and redeems a sinful life. That is how the book should end. I finally see that the real victory in my life is the privilege of serving the one who saved me, the worst of all sinners. I get to work for Jesus to bring others to him.

With that realization, my heart sings, "Thank you, Jesus, for helping me to see you. Thank you, Jesus, for never giving up on me. Thank you, Jesus, for the person reading this book right now; you have not given up on them either. Please draw the person reading this passage to you right now."

Life is a long journey. The only one who knows what's ahead on that journey is Jesus. I have failed miserably in my life when I got ahead of the Lord. You might do the same thing at times. Maybe it's in your character to grab the bull by the horns and do whatever it takes to get the job done. If that is your natural tendency, then how do you learn to be still and listen before the Lord?

Fortunately, God knows your heart and your desires. He knows what it takes to force you to slow down. But I have found that when God sees you taking even small measures to slow down to listen to him, he will move mountains for you. Yes, he will! I don't know where you are on this journey called life, but I pray that after reading what God has done in my life, you will turn your life over to the only one who can save, the only one who can restore, the only one who can show you how to love.

As you can see from my life, God used many things to get my attention and to help me see my need for him. I know if you were to look back over your life right now, you would see that God has had his hand on everything that has happened to you. He has protected you and loved you in ways you could never imagine. But above all else, one thing is for sure: he will use you for his perfect plan and purpose.

One of the great stories in the Bible concerning the *why*s of life is the story of a man born blind. The disciples asked Jesus who had sinned to cause the blindness: the boy or his parents. Many people mistakenly assumed that if something bad happened, then someone must have sinned. I find great comfort in the passage where Jesus talks about this man born blind:

> As he went along, he saw a man blind from birth. His disciples asked him, "Rabbi, who sinned, this man or his parents, that he was born blind?"
>
> "Neither this man nor his parents sinned," said Jesus, "but this happened so that the work of God might be displayed in his life."
>
> —John 9:1–3

Everything that has happened in your life will eventually reveal the glory of God, if you allow it to. What a great passage of hope to get you up off the ground and to help you keep on keeping on!

When you understand that all things work in accordance with God's perfect purpose, the things of this world that once seemed important will soon fade away. And when that happens, Jesus will become clearer. Open your hand today and give to the Lord *all* your worries, concerns, and hope for the future. Yes, God has a perfect plan just for you.

If you have lost something of great importance to you, please don't fret. God will replace it with something better. I never got my flying career back, but God gave me something better: he gave me eternal life. There is a destiny for each one of us. There is a plan for each one of us. There is a hope and a future for each one of us.

Don't ever be afraid to fail. The Lord has given you talents that you need to use. Failing is *not* trying, but God rewards those who try. Remember, failing doesn't mean you are a failure. It just means that you are one step closer to success.

I pray that the story of my life will not draw you to what I have done, but to what God has done in and through me. When we can serve without expecting anything in return, then that is true success.

Look upward, and keep getting up when you are knocked down. In all things give praise to Jesus. He is worthy of your praise. Now go out in the Lord's power and make a difference for the only one who can save, the only one who can give hope, the only one who can bring back your joy.

As you can see from this book, *nothing* you have done in this life has been a failure. Oh, by the world's standards, maybe you've failed, but not from God's view. Everything—and I mean everything—that has happened in your life has been ordained by God for his, not your, perfect purpose. He has a plan for you, and the most important part of that plan is for you to believe in him so that you will not only spend eternity in heaven but also have joy here on earth.

The Biggest Decision of Your Life

I have waited until the very end of this book to share with you how you can know for sure that you are going to heaven. I pray that you have believed everything I have shared with you about my life and realize that on my own I could never have written this book. If left to myself, I cannot put two sentences together. But God can! So the reason I have written this book comes down to these next few paragraphs. This is the most important thing you will ever read in your life.

As I've said before, Evangelism Explosion is a great course that explains the gospel in a thirteen-week study. Approximately seventeen years ago, I took the course, and it helped me not only to have the tools to share the gospel of Jesus but also to understand my own faith. Some of the thoughts and ideas below are gleaned from that course.

The Bible was written so that we may know how to inherit eternal life. As 1 John 5:13 says, "I write these things to you who believe in the name of the Son of God so that you may know that you have eternal life."

If you were to die today and God were to ask you, "Why should I let you into my heaven?" what would you say? Over the years, I would have to say that 85 percent of the answers I have gotten are some version of the following: "I have tried to be a good person"; "I have tried to be a good parent"; "I have tried to go to church and teach Sunday school"; or "I have tried to give money to the poor." The key words in all these answer are "I have tried." Instead of telling you that answer is wrong, I am going to tell you I have good news for you, and this it: heaven is a free gift.

Romans 6:23 says, "For the wages of sin is death, but the gift of God is eternal life through Jesus Christ our Lord." Now if someone were to give you a gift and you tried to pay even a penny for it, then it would not be a gift anymore, would it? Of course not. But in this world, we usually associate having good things with working hard for them or sacrificing in order to obtain them. It is hard to wrap our minds around the fact that heaven is free. Well, it is free to us, but it did cost someone his life.

A lot of people in this life try to do things to put themselves into a favorable position with Jesus. *If I do _________, then Jesus will love me,* they think. Or, *My life has been full of so many trials that God will take pity on me*, they conclude. But Ephesians 2:8–9 clearly says, "For it is by grace that you have been saved, through faith—and this is not from yourselves, it is the gift of God—not by works, so that no one can boast."

From this we can see that God's grace is what gets us into heaven, but let's take a look at what exactly grace is. God says that no number of good works will get us into heaven. God does not want anyone to boast because of his works. You may be protesting, "If I have done all these good things over the years but they don't count, then how will I get into heaven?"

First of all, you need to realize that everyone on this earth is a sinner. Romans 3:23 says, "For all have sinned and have fallen short of the glory of God." Sin is any word, thought, deed, or attitude that falls short of the glory of God. So if we all fall short and no sin is allowed in heaven, then obviously we are all in deep trouble.

Exodus 34:7 declares, "Yet he does not leave the guilty unpunished." Since God is a holy and just God, he *must* punish sin.

This creates a huge dilemma. Heaven and God are perfect; you and I are not. Let's suppose that you commit only five sins a day. But since sin is any thought or deed that falls short of God's glory, by the time you get to work in the morning, you may have already committed ten sins or even more. By the end of the day, you may have a hundred sins accountable to you. But for the sake of our example, let's just use five sins a day.

Five sins a day times a year equals more than 1,825 sins a year. Over a lifetime of seventy-five years, you would commit well over 100,000 sins. If you stood before God with this record, knowing that *not even one sin* is allowed into heaven, then yikes! God would have to look at not one page of sins, but an entire book, and you would be finished.

So God saw that you and I were dead in our sins and in desperate need of a savior. There is no way that we can get into heaven based on what we do or who we are. In order to save mankind, God needed a perfect sacrifice, one without sin, to pay for the sins of all the rest of us. Jesus Christ, born of a virgin, was God in the flesh, and he came to save sinful man.

John 1:1, 14 says, "In the beginning was the Word, and the Word was with God, and the Word was God. . . . The Word became flesh and made his dwelling among us. . . . We have seen his glory, the glory of the one and only Son, who came from the Father, full of grace and truth." Amen to God's mercy to send Jesus to this earth! Jesus lived with us, *but he did not sin*. Because he lived on this earth as we do, he knows our pain, understands our temptations, and relates to our hurts.

In the Old Testament, unblemished animals were sacrificed, and their blood atoned for the people's sins. But the priests had to keep offering the sacrifices on behalf of the people over and over and over again. However, Jesus Christ, the Lamb of God, was the perfect sacrifice for our sins. He willingly submitted himself to insults, allowed himself to be spat upon, and let himself be hung on a cross for you and for me. We are the ones who deserved to be crucified—not Jesus. But God needed a perfect sacrifice to save mankind. God needed Jesus.

John 19:30 records, "When he had received the drink, Jesus said, 'It is finished.' With that, he bowed his head and gave up his spirit." Yes, Jesus paid the price, and Satan was rejoicing. Jesus was dead. Many people mistakenly thought that the work of Jesus was finished and that there was nothing more he could do. But death could not hold Jesus. On the third day, God raised him from the dead. Jesus conquered death not only so we would have forgiveness for our sins—and it is sin that keeps us out of heaven—but also so we would have him to take us to heaven.

But how can we have the assurance that heaven is in our future? There are three types of faith, but only one will work to get us into heaven. First, there is a temporal faith. With this kind of faith, we may pray to God to get over a sickness or pass a test, but after the crisis is over, we forget about God and go back to business as usual.

Then there is an intellectual faith. A lot of people believe in Jesus in the same way they believe in George Washington or Abraham Lincoln. They believe that he lived and died, but they don't believe that he has any power. James 2:19 speaks of this kind of faith: "You believe that there is one God. Good! Even the demons believe that and shudder." You see, Satan and his demons believe in the person of Jesus, but they are certainly not going to heaven. Obviously, intellectual belief in Jesus is not the kind of faith God is looking for.

The only faith that will get you into heaven is saving faith. What is saving faith? Since I love to fly, let me use that for an analogy. Let's say you are flying from New York to Los Angeles and are waiting in the terminal for the boarding announcement. The plane is more than capable of carrying you to Los

Angeles, but you first must get up from where you are and board it. Even though you look at the plane and know in your heart that it can take you to Los Angeles, you have to get up and take action. You have to walk down that Jetway and take your seat on the plane.

Acts 2:21 promises, "And everyone who calls on the name of the Lord will be saved." The great news is that *everyone* can inherit eternal life with Jesus. Everyone who exercises saving grace to call on the name of the Lord will gain entrance to heaven. No number of good works will get a person into heaven—only saving grace. Some of you may be asking, "Then why do good works?" We do good works because of what God has done for us, not because of what we can do for him. There is *nothing* we can do to get into better standing with our creator. We will be joyful believers when we do all things out of thankfulness instead of duty.

Remember, it is God who calls us to be saved. You may be reading this right now and your heart is not stirred to make a commitment. Three weeks before I was saved in 1992, if you had given me a Bible or mentioned the name of Jesus, I would have sworn at you and thrown the Bible back in your face. I say this horrible statement to make the point that it is only God who changes our hearts. We don't play games with God by saying, "Well, God, I am going to do you a favor and sign up for this eternal-life thing." No, it is God who brings us to repentance and a loving desire to know and trust him. If your heart is not stirred as you read this section of the book, then please pray that it would be.

So here is the moment of truth. Do you believe that Jesus came to this earth, was God in the flesh, lived a sinless life, died on the cross, and then rose from the dead to pay the penalty for your sins and to purchase a place for you in heaven? If so, then if you pray the prayer below and it expresses the desire of your heart, then you will be saved for eternity.

But also remember, it is God who calls you; you do not call him. God is the one who changes your heart. If your heart has been stirred as you've read this book, then God did that. He is bringing *you* unto himself. If you had no stirring about the truth of the gospel, then no prayer will get you into heaven. It is only the prayer of a repentant heart that sees its need for a savior that counts. If that is your desire, please pray this prayer:

> Lord Jesus, I am a sinner and finally realize that my works will not get me into heaven. Forgive me for my sins, for trying to be the captain of my destiny and trying to control my future. It seems that everything I have done has been futile, and my joy has been incomplete. I know, Jesus, that you died on the cross and rose from the dead to pay the penalty for my sins. Jesus, please come into my life now and take over. I desperately need your help. I turn my life over to you and trust in you, Jesus, for my salvation. Jesus, thank you for saving me. Please help me to get to know you more each and every day. Amen.

If you prayed this prayer and it expressed your heart's desire, then you are now a child of the King with an inheritance in heaven. Nothing can take that away from you. Satan may throw roadblocks in your path, and he may try to stop you from sharing this good news with others, but he cannot take away your salvation. Now that you have prayed this prayer, Jesus has sent his Holy Spirit to live inside of you forever. Amen!

When you read the Bible now, it will come alive; and when you pray, you will experience promptings to do certain things. Your life now will have new meaning. If you have made this decision for Christ, please e-mail me at brad@theultimatedecision.com so I can send you some literature to help you grow in your faith.

The decision to follow Christ is the most important decision that you could ever make in your life. Rejoice in the love of God that he sacrificed his Son so that you could have eternal life. That is the good news of the gospel of Jesus Christ. Amen to God’s love, and amen to those of you whose names are now written in the Lamb’s Book of Life!

I have enjoyed sharing my life with you. Now it is time for you to share your life with others. God makes no mistakes, and he creates everyone for a purpose. Go and tell others about what God has done in your life. If you are just getting started or if you have failed but want to try again, great! God is in the restoration and forgiveness business. Just keep on keeping on, and I will see you on the other side!

Thirty-Day Action Plan

Individual and Group Study

Chapter 1: Flight 1992

Action Plan

Psalm 41:2: "The Lord will protect him and preserve his life; he will bless him in the land and not surrender him to the desire of his foes."

It is apparent that the devil wants you and me for his purposes. Maybe you have felt that someone or something was against you, an unforeseen force. No matter how painful or questionable the attacks you are suffering, know that the Lord will protect you when you call to him.

What can you do right now to throw off the world's concerns and focus on the Lord, the creator and protector of your life?

____

____

____

____

____

____

____

Chapter 2: Good Times

Action Plan

Ecclesiastes 11:9: "Be happy, young man, while you are young, and let your heart give you joy in the days of your youth."

Some of you are older, and your youth has passed you by. Are there events from your past that you have trouble letting go of? If so, what are they?

____

____

____

____

____

____

In order to live a Spirit-filled life directed by the Lord, you need to let this event or events go. Let God have them, and then pick them up no more. You can do this only in the Lord's strength. Give the events over to the Lord right now.

Pray now for forgiveness from harboring resentment, and God will set you free. Remember, God forgave us when we were deep in our sin. We cannot accept forgiveness from God for our sins and then refuse to forgive others. Forgive *all* that has been done to you, and the Lord will set your heart free. Write today's date, ________________________, as the day you let it go. Remember to never pick it back up again.

Chapter 3: Gates of Hell Opened

Action Plan

2 Corinthians 5:1–5: "Now we know that if the earthly tent we live in is destroyed, we have a building from God, an eternal

house in heaven, not built from human hands. Meanwhile we groan, longing to be clothed with our heavenly dwelling, because when we are clothed, we will not be found naked. For while we are in this tent, we groan and are burdened, because we do not wish to be unclothed but to be clothed with our heavenly dwelling, so that what is mortal may be swallowed up by life. Now it is God who has made us for this very purpose and has given us the Spirit as a deposit, guaranteeing what is to come."

Trouble does seem to come when we least expect it. After many trials, we can sometimes live in fear, always expecting the worst. We lose our joy. But God tells us that in him we can have peace.

How could you allow God instead of the things of the world to be your portion in life?

____

____

____

____

____

____

Could you set aside some quiet time to pray and read God's Word? That is what your hope is in. When you have hope in the Lord, then his peace will follow. Trials and tragedies will come; do not be surprised by that. But the one who created all things is able to get you through your darkest hour.

Chapter 4: A Gift of God Revealed

Action Plan

1 Peter 4:10: "Each one should use whatever gift he has received to serve others, faithfully administering God's grace in its various forms."

God has given you at least one gift or even more. As we see in the above verse, the purpose of your gift is for you to serve others. I used the God-given gift of running to serve my own self-interests, and it consumed and destroyed me.

Many try to serve others, but their hearts are not in it. We need to serve others with joy, knowing that our gifts are not ours. Our gifts were given to us so that we could serve others. If we do not use our gifts properly, the Lord may take them from us and give them to someone else.

How could you use the gifts God has given you to joyfully serve others instead of serving yourself?

____

____

____

____

____

____

Chapter 5: Where Is My Security?

Action Plan

John 16:33: "I have told you these things, so that in me you may have peace."

Most of us put our security in things. We think that when the next promotion comes, when we earn a bigger paycheck, when we get married, or when some other future type of event occurs, then everything will be right. The truth is that the Lord leaves a place in our hearts that only he can fill. But many of us certainly try to fill it with stuff instead of with God, don't we?

What have you put in your life ahead of God? Be honest!

____

____

____
____
____
____

Now that you know what it is, what steps will you take to put God first?

____
____
____
____
____
____

Do you really believe that you can have peace in a troubled world? If your answer is no, then I want you to know that there really are steps you can take to find peace and joy in Jesus. But any relationship takes time. If you are married or have a best friend, that relationship did not just happen. It took time, commitment, and selfless acts of truly setting aside your own agenda. So is it in your relationship with Jesus.

Could you commit to getting up fifteen minutes early every morning? I would suggest you do this and start reading the book of John in the New Testament. You can start with a few verses or a chapter a day. After you have read the passage, pray for a few moments. God already knows what is on your mind, so just ask him for help.

We sometimes make prayer tough because we think it has to be eloquent, but some of my prayers have been as short as two words: "Help me!" When you take just two steps towards Jesus, he takes twenty steps towards you. Please make that commitment to pray today.

Chapter 6: I'm Shot

Action Plan

Psalm 139:16: "Your eyes saw my unformed body. All the days ordained for me were written in your book before one of them came to be."

Since God knows everything before it happens, should we worry at all about anything?

____

____

____

____

____

____

God works everything out—even being shot!—for his perfect purpose. Is there something that you are worried about right now that you need to surrender to the Lord? Please be honest; it is part of the healing process to identify it and write it down.

____

____

____

____

____

____

In the Lord's strength, let it go. *Never* pick it up again!

Chapter 7: Watching Life and Death

Action Plan

Romans 8:24–25: "For in this hope we were saved. But hope that is seen is no hope at all. Who hopes for what he already has? But if we hope for what we do not yet have, we wait for it patiently."

As you look back over your life, can you identify an instance in which you know that God protected you from harm?

____
____
____
____
____
____

Why not allow the memory of this divine intervention to give you newfound hope that God loves and protects you? What could you do today to have joy in your life?

____
____
____
____
____
____

You can think of only one thing at a time. If you have trained your mind to think negative, worrisome thoughts, could you try to stop today? When those thoughts creep in, give them over to Jesus. He is more than able to conquer any fear you might have.

Chapter 8: Terror

Action Plan

Proverbs 3:25: "Have no fear of sudden disaster or the ruin that overtakes the wicked."

Many of us worry about things that never happen. We lose our joy when we do this instead of trusting in God to keep his promises. Ouch! A lot of us have terror in our hearts because we are trying to make this world heaven. We need to trust in God fully, and then we will have joy and hope for our inheritance—our future home in heaven.

What is holding you back from desiring heaven more than the things here on earth?

____

____

____

____

____

____

Whatever you have written down is probably the thing you are putting ahead of God. Find two verses in the Bible that talk about God's promises and his love for you. Write them down now and carry them with you.

____

____

____

____

____

____

Chapter 9: Dreams Become Reality

Action Plan

Jeremiah 29:11: "For I know the plans I have for you," declares the Lord, "plans to prosper you and not to harm you, plans to give you hope and a future."

Has God ever given you prosperity, but you squandered it?

____
____
____
____
____

How can we guard against thinking that our gifts come from what we have done for ourselves instead of from God?

____
____
____
____
____
____

It is easy to cry out to God when everything is crumbling around us. What could you do to guard against putting God second and third in your life when things are going great and crying out to him only when things begin to fall apart?

____
____
____
____
____
____

Chapter 10: Are You Kidding Me?

Action Plan

1 Peter 4:12–13: "Dear friends, do not be surprised at the painful trial you are suffering, as though something strange were happening to you. But rejoice that you participate in the sufferings of Christ, so that you may be overjoyed when his glory is revealed."

We have all suffered setbacks when we least expected it. When that happens, we can become bitter and hold on to resentment, or we can choose to let it draw us closer to Jesus. How could you allow your current trial to help you get better rather than grow bitter?

____
____
____
____
____
____

Maybe you are not a believer in Jesus and are at the point in your life where I was in this chapter. As you can see from my story, because of God's love for you, he will work out all things in your life in accordance with his perfect plan.

____
____

Is there something in your life that makes you feel as if you have nowhere to turn? What is it? Will you finally ask God into your life and let him take control?

____
____
____
____
____
____

If not, what is holding you back?

____
____
____
____
____
____

Chapter 11: Hired

Action Plan

Ecclesiastes 5:19: "Moreover, when God gives any man wealth and possessions, and enables him to enjoy them, to accept his lot in life and be happy in his work, this is a gift of God."

There is nothing worse than working at a job you hate. But as you can see from the above passage, it is only God who allows you to be happy in what you do. No matter where you are in your life, it isn't by chance.

How could you be joyful in what you are doing today?

____

____

____

____

____

____

If you feel God is leading you to another job, pray for his wisdom, strength, and clarity. Then, like Peter, get both of your feet out of the boat. After all, this life was meant to be lived.

Many of us, however, have a hard time just existing, much less truly living. Does that describe you? If so, how could you stop just existing and start living the life God intended for you?

____

____

____

____

____

____

Chapter 12: Flight 1992, Continued

Action Plan

Hebrews 1:14: "Are not all angels ministering spirits sent to serve those who will inherit eternal salvation?"

As you can see from the story in this chapter about ice on the aircraft, this airplane should not have stayed airborne. Sometimes there is no explanation but a supernatural one. God has ministering spirits that he sometimes releases on our behalves. Have you ever seen an angel or felt that you experienced divine intervention? Write about it in the space below.

____

____

____

____

____

____

Though we may not know it, we are constantly in a war taking place in an unseen dimension. Angels and demons are fighting for our eternal souls. After we become Christians, demons continue to fight us, trying to keep us from ministering to others. Do you view spiritual warfare as something real or something imagined? Explain your answer.

____

____

____

____

____

____

If you consider spiritual warfare as real, what role do you think prayer plays in it?

____

____

____

____

____

____

Chapter 13: Now What?

Action Plan

Ecclesiastes 2:10: "I denied myself nothing my eyes desired; I refused my heart no pleasure. My heart took delight in all my work, and this was the reward for all my labor."

As you can see from the story in this chapter, I tried many different things in an effort to find fulfillment. One way I tried to find happiness was through sexual encounters. I was not saved at the time and had no conviction against this behavior. Is there something you are currently involved in that you wish you were convicted of in your spirit? Write about it below.

____

____

____

____

____

____

Chapter 14: Flying the Line

Action Plan

James 1:2–4: "Consider it pure joy, my brothers, whenever you face trials of many kinds, because you know that the testing of your faith develops perseverance. Perseverance must finish its work so that you may be mature and complete, not lacking anything."

I was finally fulfilling my life’s dream when all of a sudden my flying career was taken away. When we face tough trials in life, we can become either better or bitter. It is tough becoming better when we focus on the past. But the trials of life that last a long time provide the perfect opportunity for us to draw closer to God.

What has been the toughest trial in your life?

____

____

____

____

____

____

Looking back, can you see that this trial, while it may not have been pleasant at the time, did develop perseverance that made you grow in your faith? Write about your experience in the space below.

____

____

____

____

____

____

The next time a trial comes your way, will you be able to handle it any differently from the way you have handled trials in the past? What will be different this time?

____

____

____

____

____

____

Chapter 15: A New Normal

Action Plan

Proverbs 3:5: "Trust in the Lord with all your heart and lean not on your own understanding."

What are you doing in your life right now that doesn't make sense?

____

____

____

____

____

____

Does it not make sense because you are leaning on your own understanding, or does it not make sense because you are relying on God? Either way, you will never be able to figure this life out.

What are some steps you could take to trust in the Lord starting now?

____

____

____

____

____

____

Chapter 16: Panama City

Action Plan

Proverbs 16:9: "In his heart a man plans his course, but the Lord determines his steps."

Have you ever thought your plan was foolproof only to have everything in it turn upside down. If so, what happened?

____
____
____
____
____
____

Many times in life, we make plans and then ask the Lord to bless them instead of asking the Lord first what our plans should be. I was prepared physically for the swim at Panama City, but I wasn't prepared for the riptide. Sometimes life throws us a curveball when we are expecting a fastball. How do we switch gears without worrying?

____
____
____
____
____

What could you do in the future to make sure your plans align with God's plans?

____
____
____
____
____
____

Chapter 17: Heaven's Destiny

Action Plan

Ephesians 1:11: "In him we were also chosen, having been predestined according to the plan of him who works out everything in conformity with the purpose of his will."

In my life, there was a time and place determined beforehand for me to accept the free gift of God's salvation. If you have accepted the Lord as your Savior, do you still have that joy as when you first believed? If not, what has happened?

____

____

____

____

____

____

What steps could you take to get back that excitement and peace in your spirit?

____

____

____

____

____

____

Chapter 18: A Change Of Heart

Action Plan

Philippians 4:12,13 "I know what it is to be in need, and I know what it is to have plenty. I have learned the secret of being content in any and every situation, whether well fed or hungry, whether living in plenty or in want. I can do everything through him who gives me strength."

No matter what happens in your business life, it is the Lord who allows you to enjoy your work. Have you been doing things backwards? Have you been thinking that when you get to a certain point or place, then you will be happy and or content?

If so, write about it below.

I have found that there is never a perfect time to start anything or to be happy. If we are waiting for something to happen before we do something else, we will never do anything. There will always be something to make us discontented.

What could you do to live a life of joyful expectancy that God will do great things?

Chapter 19: Hope

Action Plan

Jeremiah 29:11: "For I know the plans I have for you," declares the Lord, "plans to prosper you and not to harm you, plans to give you hope and a future."

Have you ever failed so many times that you lost hope? Can you recall a time when this happened? Write about it in the space below.

What could you do right now to know that the Lord is out "to prosper you and not to harm you"?

____

____

____

____

____

____

Are you ready to let go of the past and focus on the future? Would you pray that the Lord would let you experience victory rather than defeat, hope instead of despair? If so, please ask Jesus right now to help you do this.

____

____

____

____

____

____

Chapter 20: Will Business Ever Change?

Action Plan

Ecclesiastes 1:7: "All streams flow into the sea, yet the sea is never full. To the place the streams come from, there they return again."

I love this verse. Ecclesiastes was written by King Solomon, the son of King David. Solomon had everything any man could ever want. At the end of his life, however, he recognized that everything was futile if the Lord was not in it.

The analogy of the streams flowing into the ocean without the ocean ever becoming full describes our working from morning till night, day after day, week after week, and year after year.

But in the end, nothing we do will make a dent in the larger scheme of things. There will always be more work to do.

What is it that you feel is so important that you are sacrificing time with your family in order to achieve this goal?

Is it worth it? If not, how or what will you change moving forward?

Chapter 21: A Life-Changer

Action Plan

Jeremiah 1:5: "Before I formed you in the womb I knew you; before you were born I set you apart."

Psalm 27:3 tells us that children are a gift from the Lord. If you are married and trying to have children or if you are not married yet but want a family, please do not give up hope. If you do have children and at times wonder if you did something wrong that made you have the children you have, stop thinking that. God made your children, and he made them perfect for his purposes. Does this statement help you view your children any differently? Write about it below.

Raising children can be a huge blessing and a huge challenge all at the same time and even in the same day! Do not fret about anything, as God knows your situation. Call out to the Lord for his wisdom in raising the children he created and gave to you. Remember, you have your children at home only for a very brief time, and then they are off. Do the best that you can today, but give your children to the Lord. Do not put them above God or love them more than God, which is easy to do. Does the thought of putting God first above your children hurt, or does it give your mind some rest? Explain your answer in the space below.

God urges us not to worry about anything, but many times we worry about our children. Are you able to let things go and let the one who created your children take over? Believe this: God loves your children more than you as a parent ever could, so stop fretting about them. Let go and enjoy them.

Chapter 22: NASCAR 101

Action Plan

Proverbs 16:3: "Commit to the Lord whatever you do, and your plans will succeed."

Our business started to crumble for a number of reasons. One of the reasons was that we were not daily seeking the wise counsel of others, especially the Lord.

____

A lot of times in business, you can start off great by praying and giving the day to the Lord, but after a while, when things become hectic, you can easily forget what got you there. When our business began slumping, we just put our heads down and kept plowing forward. Though we were working hard, we were not working unto the Lord.

____

____

Is there something that you are doing right now that is not working for you even though you are working hard? What is it?

____

____

____

____

____

If you are struggling in a business, pray for a couple of godly men or women to take a look at your business for you. Turn yourself over to their wisdom, and listen to what they have to say. Whom will you ask? Now call them, but first write down what you will discuss with them.

____

____

____

____

____

Chapter 23: Panama City Failure to Full Ironman

Action Plan

Joshua 1:7: "Be strong and very courageous. Be careful to obey all the law my servant Moses gave you; do not turn from it to the right or to the left, that you may be successful wherever you go."

The Ironman is a great metaphor for life. The day is so long and the distance is so daunting that you cannot think about the entire race at once. When you are in the swim, you need to think only about your next stroke, not that you have 112 miles to go on the bike and then a marathon to run.

In life we can think too far down the road: *What if my kids don't turn out right? What if I lose my job? What if the test results come back positive?* and on and on. What could you do today to stop the *what if*s of life and learn to live only in the moment?

____
____
____
____
____

There were many times that I thought about quitting, as it was such a difficult race. The sign "Pain lasts for a moment, but quitting lasts forever" helped me keep going. Do those words mean anything to you in your current struggle?

____
____
____
____
____

Running across the finish line with my family after twelve hours of continuous exercise was a very spiritual moment. When we get to heaven, we all want to hear, "Well done, good and faithful servant." What could you do today to keep the finish line of life in sight while also living for today?

____
____
____

Chapter 24: Ironman Wisconsin

Action Plan

Philippians 3:12-14 "Not that I have already obtained all this, or have already been made perfect, but I press on to take hold of that for which Christ Jesus took hold of me. Brothers, I do not consider myself yet to have taken hold of it. But one thing I do: Forgetting what is behind and straining toward what is ahead, I press on toward the goal to win the prize for which God has called me heavenward in Christ Jesus.

I thought I had trained hard enough for the Ironman event in Wisconsin, but sometimes in life, circumstances beyond our control arise. God, however, is still in control, orchestrating everything in this world and in the heavens.

What is the toughest thing God has ever allowed you to experience?

This book is about getting back up after you have failed. You may have experienced a great failure, and it is making you feel unworthy. You may have lost your confidence to carry on. What can you do today to stop this vicious cycle and forget the past, knowing that God has set for you a glorious future?

____

____

____

Chapter 25: Beginning of the End

Action Plan

2 Corinthians 4:8–9: "We are hard pressed on every side, but not crushed; perplexed, but not in despair; persecuted, but not abandoned; struck down, but not destroyed."

Has there been a time in your life when you have felt that you just couldn't go on? Write about it in the space below.

____

____

____

____

____

____

How can God's Word and his promises help you to keep on keeping on even when everything else tells you to stop?

____

____

____

____

____

____

Chapter 26: Full-Time Ministry

Action Plan

1 Thessalonians 5:18: "Give thanks in all circumstances, for this is God's will for you in Christ Jesus."

After the failure of almost everything I did, I found myself asking, "God, what do you want me to do?" Ever been there? But when I look back over my life now, I can see that everything I have done has prepared me for today. I would like to say that my faith brought me into full-time ministry, but it was really God's circumstances that did.

Are you happy in your work? Explain.

____
____
____
____

If you are not happy, then why not?

____
____
____
____
____
____

Is there something you could do to change careers or change the way you view your job and the people you work with? Life is too short to be miserable.

____
____
____
____
____
____

Chapter 27: Life's Storms

Action Plan

Matthew 7:24–25: "Therefore everyone who hears these words of mine and puts them into practice is like a wise man who built his house on the rock. The rain came down, the streams rose, and the winds blew and beat against that house; yet it did not fall, because it had its foundation on the rock."

Chase's autism has been a huge blessing and a huge challenge all at the same time. Julie and I would never have been able to get through this without Jesus and the Holy Spirit giving us hope.

After all the defeats in my life, I realize that everything I tried to build without God was built on sand, and it eventually crumbled. Is there something that you are putting your time and effort into that will crumble when a little wind and rain comes? If so, write about it in the space below.

____
____
____
____
____
____

How could you change your mind-set so as to build your dreams upon the Rock?

____
____
____
____
____
____

Chapter 28: Natural Stick

Action Plan

Proverbs 13:12: "Hope deferred makes the heart sick, but a longing fulfilled is a tree of life."

After a twenty-seven-year wait, it looked like my longing to be an airline pilot again was going to happen. Do you have a dream that has not come to pass or a talent from God that you are not using?

____

____

____

____

____

____

What steps could you take to make your dream a reality?

____

____

____

____

____

____

Chapter 29: Redemption

Action Plan

2 Corinthians 12:7–10: "To keep me from becoming conceited because of these surpassingly great revelations, there was given me a thorn in my flesh, a messenger of Satan, to torment me. Three times I pleaded with the Lord to take it away from me. But he said to me, 'My grace is sufficient for you, for my power is made perfect in weakness.' Therefore I will boast all the more gladly about my weaknesses, so that Christ's power may rest on me. That is why, for Christ's sake, I delight in weaknesses, in insults, in hardships, in persecutions, in difficulties. For when I am weak, then I am strong."

God has given me so much, but most of the time, I have been so ungrateful. I have had a very hard time accepting God's grace because in this life it seems that you have to work hard for anything that's worth something. My life has been one setback after the next. The loss of my flying job *again* would have done me in if I had not known that the Lord was directing my life. I finally realized that he has a perfect plan for my life. The question is, do you really know that God has a perfect plan for you also? Write your thoughts in the space below.

____

____

____

____

____

____

If you do not know that God has a perfect plan for you, then you will not find fulfillment in anything you do. If you are suffering trial upon trial and failure upon failure, know that God is at work in your life, not to harm you, but to set you free. Until you suffer the pain of loss, you will not experience the victory of grace. Remember, this earth is not your real home. Pray right now that you would not see failure as defeat, but that you would see failure as bringing you one step closer to seeing God's glory and one step closer to seeing God's plan revealed.

Chapter 30: Will I Get to Fly Again?

Action Plan

Isaiah 55:9 ""As the heavens are higher than the earth, so are my ways higher than your ways and my thoughts than your thoughts."

Have you ever wanted something so badly that you could taste it? If so, what was it?

____

____

Then, when that dream was taken away, did you get mad at God, or were you able to trust in God completely and not lean on your own understanding?

If you are still mad at God for taking away your dream, what would help you understand that God has something better for you?

Chapter 31: God’s Perfect Plan

Action Plan

Hebrews 6:10–12: “God is not unjust; he will not forget your work and the love you have shown him as you have helped his people and continue to help them. We want each of you to show this same diligence to the very end, in order to make your hope sure. We do not want you to become lazy, but to imitate those who through faith and patience inherit what has been promised.”

God's plan is perfect, but sometimes we can feel as though we are running in place, not getting anything done. Has this ever happened to you? Write about it below.

____
____
____
____
____

The above verse admonishes us not to be lazy. We know God has given us talents, and we know we need to do the best we can with these talents. Though we may know all this in our heads, how do we get this into our hearts?

____
____
____
____
____

What could you do to help you follow God's perfect plan instead of making decisions first and asking God for his blessing afterwards?

____
____
____
____
____
____

Chapter 32: The Biggest Decision of Your Life

Action Plan

John 3:16: "For God so loved the world that he gave his one and only Son, that whoever believes in him shall not perish but have eternal life."

We can read this verse many times and just gloss over it after a while. I have often mistakenly thought my failures were a result of God's being mad at me, and I have often struggled to grasp God's love for me. In the back of my mind, I always had a performance clause that motivated me. I knew my life in Christ was based on God's grace, *but* I always had to mix in a little bit of works with it.

God loved you and me from the beginning of time, even though there was nothing in us to merit saving. He sent his Son to die in my place and yours. It is you and I who deserved to hang on that cross where Jesus hung. But God never brought up our failures; he only brought up the way to eternal life. God didn't keep pounding us with the wickedness of our sin, but he kept offering his glorious grace. He didn't keep loading us up with burdens, but kept showering us with love. That is the God we serve. That is the only God who saves, and that is our God.

Do you know God? This chapter tells you how you can know for sure if you are going to heaven and sharing eternal life with Jesus. There is not a more important decision in this world. What have you decided? Will you follow Jesus and put your faith and trust in him? Write your decision in the space below.

____

____

____

____

____

____

If you made a decision to follow Christ, please e-mail me at brad@theultimatedecision.com or send a note to me at the following address: The Ultimate Decision, P. O. Box 2337, Huntersville, NC 28070. I will be sure to send you information to help you grow in the faith.

Thank you for taking this journey with me!